Praise for *Your Call is Important ...* Najafov

First of all, I would like to personally thank Boris for sharing his idea about writing a book at the very early stage of this tremendous work and finally for giving me a chance to write a review for his book.

Pay attention to the title − "Your call is important for us" − how many times each of us heard this phrase when calling different companies for service inquiries or complaints. But did you ever thought about what stands behind the simple action of dialing number and getting you call answered by the right person at the right time?

The stack of technologies and processes which comprise the modern contact center can be overwhelming and this fact is actually stopping many of us from going into details in order to really understand how this works. But now, with this is book this becomes fairly easy.

Author has divided the content in several chapters and as he states, each chapter is really can be read with a cup of coffee. Additionally, this also gives a reader the chance to consume the material at his own pace, which is also very beneficial considering the fact that time has become a commodity nowadays and there is never enough of it.

I intentionally use the term "contact center" instead of "call center" − modern day customer communication involves many ways other than telephone to interact and serve the customers. After reading the book you will understand many things about these new ways and methods as good as the technology and analytics which are helping to build a better customer experience and greater customer satisfaction, which are, in my humble opinion, the key factors that make the product or the company successful today.

I will recommend this book to any person from an engineer researching his career advancement possibilities in contact center industry to a customer service representative or contact center manager looking to broaden their knowledge about using the technology for building a great customer experience in their contact centers.

—Orhan Tagizade (IT Manager, PASHA Holding)

This book appears to be a biography of the contact center industry. Boris proficiently enlightens the reader on the evolution of the contact center industry and what's the real 9 to 6 life of a professional working here. Information transfers from peripheral to the core of the industry, its technologies, terminologies, and more.

The author discusses call center technologies such as PAB, ACD, and their drawbacks that lead to innovations in today's contact center industry. I believe it is the 'abc' of call center industry that is enough to train a new joiner eliminating the need for further studies or training. Boris has very well explained the evolution of the telephony system and how call centers worked — further advancing to the workforce and quality management. Then it talks about the journey of omnichannel customer experience and the emergence of voice analysis and chatbot application. It explains how unified agent desktops, integration, and wallboards have revolutionized agents and customers' experience. From *"may I know who am I talking with' to 'Hello Mrs. Anne, hope your refrigerator is working fine"* agents' are highly empowered and their scripts have transformed drastically.

Further, it discusses the performance monitoring and analysis that's imperative to achieve exceptional customer service. Leaping into the future trends of cloud transformation, automation, AI, and knowledgebase, the book elucidates how new technologies' integration is promoting self-help services at the contact center that customers have been looking forward to. After sharing an exhaustive understanding of each topic, I am incredibly impressed with the quiz Boris has shared for reinforcement of the learnings.

This comprehensive guide caters to the needs of every individual associated with the contact center industry ranging from network engineer or sales professional to a CEO or even the customers who, at times, feel agitated when their call is put on hold.

As the CEO of NovelVox, a contact center solution company, I am highly impressed by this book and can correlate to this experience. I would recommend people to read this book, especially the ones willing to make a career in the contact center or customer experience industry.

— Amit Gandhi (CEO, NovelVox)

Acknowledgment

Writing this book would not have been possible without the help of many people who I would like to thank herewith.

I would like to thank my colleague, Samer Shafeek, for showing me that numbers give the same things different meanings; Mohammed Fahmy for his continuous trust and long term vision; Mohamed Hilmy and Zaid Shaban for the many years of healthy arguments which I am still enjoying; Mohamed Khedr for demonstrating the value of hard work and change; Orkhan Tagizade and Amit Ghandi for reading the early version and providing valuable feedback; vendors and partners ZOOM (and Liam Anderson), Verint (and Shaheen Haque), and NovelVox for providing samples of their systems' views; Art School by Alyona Bandurina and especially Teymur Aydazada for drawing all the neat diagrams you'll see in this book; and Diana Najafova for her continuous support and stopping me from giving up.

I would also like to thank the members of my very best IST Networks solutions consulting team, Faisal Akbar, Bara'a Hazaimeh, Vaibhao Khante, Baris Bicakci, Mark Onsy, Anwar Ahdab, Nada Ammar, Ahmed Abdelbaky, and Tasniem Farahat who have tremendous knowledge and experience on which I can rely.

Finally, writing a book needs a good coffee. My special thanks to "Chez Sophie-Amwaj" for good coffee and the relaxing atmosphere where the book got its final touches prior to printing.

YOUR CALL IS IMPORTANT TO US

A beginner's guide to the
contact center and customer
experience technologies.

BORIS NAJAFOV

Table of Contents

Introduction

Almost 150 years have passed since Bell's famous, "Mr. Watson, come here. I want to see you," an invention that changed the way we do everything from booking a table at a restaurant to completing a multimillion-dollar transaction on the stock market. The invention of the phone was a milestone in a line of quite a few inventions preceding it (such as the telegraph) and following it (such as the radio, television, and even the Internet). Initially intended to transmit information (be it two-way communication—such as the phone—or one-way communication—such as a radio broadcast), in today's world, "communication technology" is no longer exclusive for communication alone; we use it in other parts of our lives, such as earning and spending money, ordering goods or services, and even sharing our emotions via social networks. Communication technology has become so easy to use and so affordable that today's average seven-year-old child has greater bandwidth and more processing power than a large datacenter could have afforded only a few decades ago. And yes, the processing power to comment on a funny YouTube video is similar to the one used back then to calculate the outcome of a nuclear reaction or a mathematical model of worldwide climate change.

Companies have quickly realized (though some are quicker than others) that using communication technology improves their ability to do business, but what is even more important is the realization that communicating *with customers* is efficient over the phone (and later, the Internet). For some businesses, it goes even further—the *model of their businesses does not rely on seeing their customers in person anymore*. Instead, these companies communicate remotely with their customers.

Examples can help to illustrate this. It is a safe bet to assume that you have ordered pizza by phone or through the Internet with a mobile app at least once in your life. When doing it, you may not have even known where the pizza shop was located. Have you ordered airline tickets from a travel agent without knowing where its office was? Today, you can order a pair of running shoes, buy a million dollars' worth of Apple shares, renew your vehicle registration, and even send your girlfriend's favorite teddy bear traveling around the world, all without ever leaving your home or office. You can also be "at your office" without leaving your home, but I recommend asking your boss first. By the way, the example of the teddy bear is not a joke but a rather expensive way to get rid of an old toy, at least for some time.

If business = communication, then business growth = more communication. Very soon, a single phone is not enough, the number of calls received daily can be measured in triple digits, and people realize they can either manage this efficiently or they simply cannot grow their businesses. This was when **call centers** were invented. Simply speaking, a call center is an organized department in a company with the purpose of taking customer calls. Imagine a room with a few dozens of tables equipped with phones, and people answering hundreds of calls daily. Strictly speaking, a call center consists of three major parts: technology—such as a phone system (and maybe a rather advanced one); people—the "operators" or "agents" and their managers; and the processes—for example, shift schedules, or standard ways to greet customers to let them know their call is important to you (which is arguably the biggest lie known to mankind after the answer to the question "does this make me look fat?").

Nowadays, the term "contact center" is gaining its popularity. Contact centers are similar to call centers, but besides calls, they also process emails, chats, SMS messages, "snail" mail, and even messages received through the worldwide network of legacy, low-quality printers (yes, I mean fax machines here).

Obviously, contact centers are not that simple. They are an entire industry, and the purpose of this book is to help you, not drown you in the ocean of acronyms, concepts, and technologies that comprise the contact center.

The book is intended for people starting their careers within contact centers and the customer care industry who are trying to understand the basics. This is never easy—I can tell you this from my own experience, having come into the industry over 11 years ago. The book will be useful for:
- technical people (such as IT and network engineers) to understand how their technical knowledge can be applied in the contact center field;

- technology sales professionals to understand the specifics of the industry and the purpose of the products and solutions they sell;
- CEOs and other types of high-level management to understand "what the hell is happening there on the third floor" (that's where the contact center is located), and "on what are they going to spend this huge budget they requested last Friday?"
- anyone who really wants to know what stands behind the phrase "your call is important to us" and why they say this even when "all our agents are currently busy"; and
- my family members, so they will understand what I have been doing from nine to six in the office for all these years.

This book is the outcome of my experience as a contact center design specialist and consultant. I often had to explain concepts and technology to other people, such as the new hires in my company, partners, and customers as a part of my job, all of them with very different backgrounds and experience levels. The book is written in a way that is easy to understand that can be read either straight from the beginning to the end or used for organizational purposes. If you have already spent some time in this business—feel free to read the chapters required to cover specific gaps. The industry is rapidly changing, but my experience with top brands has made it possible (as possible as it can be) to make this vendor-neutral. Where applicable, I will try to mention different names and meanings the reader might come across in real life to make it easy.

Because no knowledge is complete without testing and applying it, I have included a small quiz at the end of each chapter. My goal is not to prepare you for an exam here but to give you the chance to test the knowledge you have learned to feel more confident with it. As contact centers are much (sometimes, too much) concerned with acronyms, a list of these are located at the end of the book for your convenience.

The book is designed so that one chapter can be read (I would say, "consumed") every day. It may be technically possible to go faster but make sure you really understand the concepts first. The snowball of understanding will grow rapidly, and you will soon find yourself feeling completely helpless, and in a depression so deep, you may even forget to feed your cat. Therefore, for the sake of cat (even if you don't have one), be reasonable as to how much of the contact center cocktail you can consume on a daily basis to avoid headaches.

With that being said, contact centers are a dynamic business, and being in them is fun. Over the last few years, they have become a relatively new trend. The contact center is understood as part of a bigger business—the customer experience industry—the field in which technology and people work together to make life easier and more pleasant for us all, no matter where we go or what we do. Leaving an Instagram post, booking cinema tickets, or searching for a nice café for an evening date—we are all a part of this industry, and understanding how it works can be quite interesting.

Enjoy the read!

Day 1: From Bell's times to today—The evolution of telephony systems

Nowadays, contact centers are complicated creatures of human genius, costing millions of dollars and handling **dozens** of customer requests **per second**. To understand how big that number is, compare it to Heathrow Airport's handling of only two to three passengers per second on average (2016).

With all its complexity, a contact center can still be broken down into its core system and set of additional modules, working together like a well-conducted orchestra. Continuing this analogy, let's begin by listening to the first violin: our telephony system, which plays the central role. To make this easier, let's start where it all began 150 years ago. Let's take a minute (maybe slightly more than a minute) to travel back in time to the end of the 19th century. Bell and Watson have constructed these weird devices they call "phones," and they have connected two of them together with a wire.

We will do a similar thing now (at least on paper) and connect two phones together. The result looks something like this:

Bell's
phone

Watson's
phone

Diagram representing the experiment carried out on June 2, 1875, by Alexander Graham Bell and his assistant, Thomas Watson.

The next logical step is to have more than two people talking to each other, so let's add another phone and connect them. Your diagram should now be something similar to this:

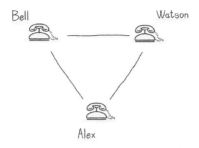

A setup with three phones connected to each other.

Though we have solved the problem for three people, it gets slightly more complicated when more phones are added:

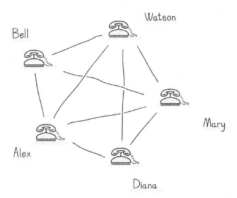

A setup with five phones connected in a full-mesh topology.

The number of wires here is hardly manageable, and it is rather obvious that we need to find a better way to connect more people. This is where the Private Branch Exchange (PBX) comes into play. The PBX is a box where all phones are connected using a *star* topology.

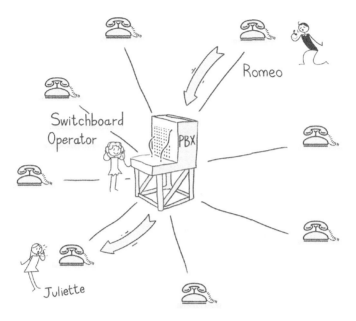

PBX unit with multiple phones connected to it using a star topology.

When one wants to make a call, he or she picks up the handset, and a person—known as *switchboard operator*—connects the phone to that of the person to which he or she wants to talk:

A PBX switchboard operator at work.

This system worked rather well, but it eventually faced limitations that had to be addressed:

- switchboard operators cannot know the names of all people, even in a small town, so "I would like to speak with Mr. Watson," may not always work well; and
- switchboard operators are human beings, and they can only handle a limited number of requests per day, thus creating the potential to bottleneck the system (humans always create bottlenecks in any system; you'll see more examples soon).

To solve these challenges, the PBX was replaced with the Private Automatic Branch Exchange (PABX), and all phones were assigned "phone numbers," unique combinations of digits distinguishing each phone from the next. Now, to reach the phone to talk to someone, you simply need to know his/her phone number.

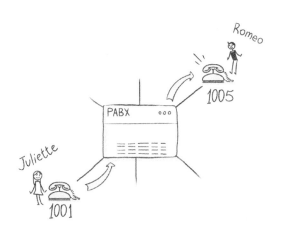

Calling a person using a phone number.

Phones also got **dials.** To "dial" a digit, one had to rotate the dial to the angle corresponding to the desired digit. When released, the dial rotated back at a constant speed, sending an electrical signal—pulses—through the wire. The PABX received and interpreted the pulses and converted them into dialed digits. You may have seen these weird phones (called **rotary phones**) in museums. You may have even used them if you—like me—were born about the same time the first episode of *Star Wars* was filmed or even before that.

A rotary phone.

A few decades later, new PABX systems appeared. Instead of sending a series of pulses on a wire, the phones sent combinations of tones, with each tone corresponding to a digit. Thus, the whole combination represented a "dialed" number. The PABX interpreted the tones to understand the number dialed. This method is called **tone dialing**, while the method discussed in the paragraph above (with weird, old phone) was called **pulse dialing**. The signals used in tone dialing (i.e., the "dialing tones") are called **dual-tone multi-frequency (DTMF) tones**. Search online if you want to know more about tone dialing and DTMF—I won't repeat the Wikipedia content here—but one thing you should note is that with tone dialing, two additional keys were introduced: the "hash" (#) and the "star" (*) keys.

Phone with tone dialing capability.

The above picture shows these two important keys on the tone keypad. The phone in the picture is similar to the phone we used at home when I was a child, but mine had an extra "S" key on the left side. Because reading manuals was considered a bad habit in the Soviet Union, the mysterious "S" key's purpose remained elusive to me until today.

The next important thing to explain here is the term "numbering plan." Communication doesn't happen exclusively inside a single company or town (or PABX). It must also reach other people connected to the telephone network. A **numbering plan** is a scheme used to assign numbers to phones—called subscribers—and defines how these phones reach each other on the connected telephony network. Let's illustrate some of the terms used to describe numbering plans using the picture below:

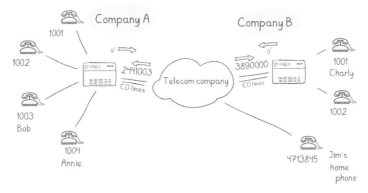

If Bob—working in Company-A—wants to call Annie, he would need to dial her **extension**, 1004.

If Annie wants to call Charly, she needs to dial an **outside line prefix**—in case of Company-A, it is the prefix 9—and then dial her global phone number, 3890000, and then extension 1001. In the last example, after dialing 3890000, Annie may be greeted by a human operator or an "auto-attendant," a robot asking her to dial the extension.

What may also happen is **direct inward dialing (DID)**, wherein each employee has both a global phone number and a local extension, somehow related or mapped to each other in a way that a person can be reached by a global phone number without the need to dial his/her extension. This can be illustrated with the same example. Imagine that Company-B signs a contract with a telecom company to own the entire range of phone numbers from 389-0000 to 389-0999. They then assign internal extensions to employees and **map** internal phone numbers to external ones. For example, subscriber number 389-0023 corresponds to internal extension 1023. External number 389-0011 corresponds to extension 1011. Such a scheme—in which an external phone number is mapped to an internal extension—is called *direct inward dialing* (DID). Using the same example from the last picture, we can say that in order to call Charly, Annie would need to dial 9 (the prefix for an outside line) and then dial Charly's DID number, 3890001.

Direct Outward Dialing (DOD) works the other way around. For example, when Charly calls Jim from his office, Jim sees Charly's phone number as 3890001 rather than the phone number assigned to his company (3890000).

It is pertinent to note that the same company may have employees with direct numbers using DID and also have some extensions without globally accessible phone numbers. For example, the phone in the common area—such as a kitchen or lobby—would not typically need a DID facility assigned to it.

Note that the **outside line prefix**, the number we have to dial to get an outside line, is not always 9. It can be 0 or any other digit or combination of digits configured by a PABX technician, though anything except for 0 or 9 may seem a bit odd. Sometimes, the PABX administrator may assign different prefixes to dial different destinations. For example, different prefixes can be set to reach mobile numbers and local phone numbers, as depicted in the example below:

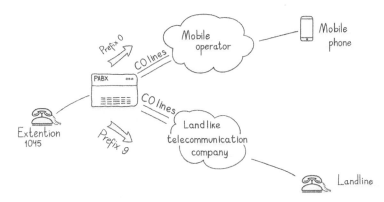

In this example, to call the mobile number, the person would have to dial the prefix 0, and to call a city landline number, he uses the prefix 9.

The lines PABX uses to connect to local telecom companies are called **Central Office (CO) lines.** This refers to the central office of the telecom company, where all lines were aggregated in the past. Having multiple CO lines for different telecom operators can reduce costs (for example, calling mobile numbers using public telephone companies may be more expensive than making calls directly through another mobile line) and have some kind of redundancy in case the telecom company (or the connection to it) is out of service for a time.

With the knowledge we have now, it is time to virtually build our first simple call center and learn about some useful PABX functions available today.

Let's imagine we own a company selling and delivering flowers, and our customers place orders by phone. We need a PABX unit and CO lines from the local telecom operator with a centralized phone number 800-124000. We have hired three operators to answer the calls and take new orders, and another two operators to answer questions regarding the status of orders. When customer Jim dials 800-124000, the **auto-attendant** plays a message saying, "Welcome to Flowers Company. To place a new order, press one. To know the status of an existing order, press two." We also need to configure two groups of operators to answer calls as **hunt groups**—each hunt group is a collection of extensions to which the PABX distributes the calls. Distributions can be **random**—each new call is distributed randomly, in a **linear** fashion. For example, the first call would go to extension 1001, the next to 1002, the third to 1003, and so on, and **the longest idle**—where the call is sent to the operator idle for the longest amount of time. Instead of a hunt group, we may have to configure the extensions in a **ring group**—all extensions call simultaneously until an operator picks up the call. Regardless of how we configure our hunt groups in the setup above, we will achieve a few *business outcomes* in our contact center, as follows:

- customers choose the service they need (by pressing one or two);
- we achieve a somewhat fair distribution of calls between operators (agents) in the same group; and
- we group our employees (agents) according to the service they provide to the customer (and therefore, according to their skills).

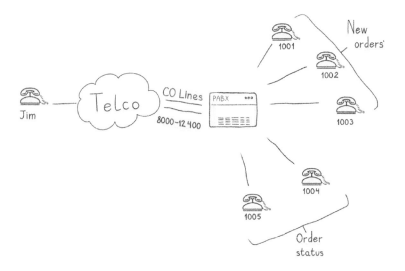

Now is the time to learn more about PABX **call control** functions, which help to serve the customers better. They are:

- call transfer, in which the PABX disconnects the call from the agent and reconnect it to another agent (or to any other extension or outside number);
- call forwarding, when calls coming to one extension are automatically redirected to another destination, such as another extension or employee's mobile phone (useful during lunchtime);
- conference, when more than two people are on the same call at the same time;
- hold, when a call is suspended—customers can listen to music on hold (MOH), a melody, or repeated message recorded on PABX—and later retrieved;
- call waiting, when one call is suspended—MOH normally plays for the customer during this time—while agents are receiving other calls.

These functions are commonly referred to as **call control** because they control where and how calls are established, answered, and ended. Remember this term—you'll need it very soon.

There are some advanced functions of the PABX that can help go a bit further toward building our call center. For an extremely small company—selling a few bouquets of flowers a day—this call center may provide adequate functionality and performance. When building something bigger and more complex, PABX functions may be too limited, even with today's most advanced PABX units. These limitations will be addressed as we move forward (I promise), but before they are, it is time to learn how *IP telephony* works.

All modern contact centers are based on IP telephony. Only their older, outdated counterparts are based on the legacy PABX units we discussed above. Don't worry—the concepts you have just learned are exactly the same and will be useful for understanding the world of IP telephony as it stands today.

To begin, let's think about a typical office workplace, containing a table with a chair, a coffee shop just around the corner on the ground floor, a PC or laptop on the table, and a phone just next to that. Compare this workplace to the one we had ten, 20, or 30 years ago, and you will find slight differences in design, particularly that monitors have become thinner and larger. What has changed dramatically is the network connection, not only the speed and reliability but the role it plays. Most of the things a typical employee does today are done on the network, from closing a deal with a new customer to submitting a vacation request, but until recently, one function that remained separate was voice (phone) communication, and companies paid a rather high price for this separation.

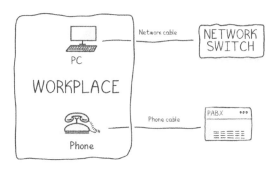

Companies had to pull and maintain two cables per workplace, one for telephony and one for the network. Accordingly, they had to manage two separate cabling systems. If an employee moved from floor one to floor five, this meant the telephony cables has to be somehow rearranged to allow the employee to maintain his extension number. In an office with 15 employees, this was still easy to do, but it became rather tricky in an office with 500 employees. It would be a nightmare for a campus with 5,000 employees. Compare this to the flexibility of a computer system in which you could easily take your laptop and move it to another desk, logon to your account, and have all of your files available to you; the inflexibility of legacy phone systems seems unforgivable.

Why don't we connect our phone systems to networks and use network connections to carry voice transmissions for us? It sounds like a plan, but a few technical challenges had to be solved:

- On a network, each device is identified with a unique IP address. These addresses are similar to 192.145.2.34 and are not very human-friendly (except for network guys who love to remember such numbers). We are used to extension numbers, such as 1004, which are much easier to remember, and with IP telephony, we still expect phones to be numbered in the same way as traditional telephony, with three- to four- digit extension numbers.

- If phones and PCs are on the same network, how does this connect to telecom networks still using old legacy connection types?
- We all know that networks are slow at times. How can we ensure that voice transmissions will have adequate quality even when the network is slow?
- Networks and hackers—as seen on TV, these two are often found together. If our phone conversations are on the network, does this mean we can be easily hacked?

Let's look at a diagram of a typical office network setup using IP telephony. This is a typical network setup with switch, router, and Internet connection, which we are used to at home and in the office. In addition to this, we have incorporated IP telephony elements here.

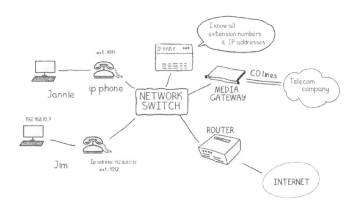

1. IP phones connect to the network, just like normal computers. To make our lives easier, most good IP phone models have two ports, so the PC can be daisy-chained to the phone, meaning you need only run **one cable per desk**. IP phones (like computers) have their unique IP addresses on the network, but users need not know these addresses.
2. IP PABX plays a role similar to that the good old PABX plays in a telephone setup. In addition to this, IP PABX maintains a list of the IP addresses of each phone and its matching extension.

Therefore, Jim will still be able to reach Jennie by dialing her extension, 1014.

3. All phone conversations pass through the network switch, which typically has something called a Quality of Service (QoS) configuration. This feature gives priority to voice traffic at all times, meaning that voices will be clear even when the network runs slower than usual.

4. Media gateways have ports to connect to networks as well as CO lines provided by local telecom companies (telco). Depending on the type of connection you get from your telco, you may need to install matching modules to your media gateway to correspond to the type of connection. For example, if your telco provides E1-PRI over copper, then you will need an E1-PRI copper module.

5. IP Phones can encrypt and decrypt voice conversations before sending them over a network. Network teams also have experience configuring many security features, such as firewalls, access lists, security policies, and other bells and whistles needed to make it secure. In all cases, this is much more secure than plain old telephony, which could be "hacked" by wiretapping the telephony cable (I will not explain it here, but google it please, if interested).

Before we finalize the chapter, there are a few more things you need to know about IP telephony. The first one is the term **protocol**. Protocol is the language two systems use to understand each other. It is the same as two people who must find a common language with which to communicate. Systems are often capable of supporting a few protocols in the same way, people can speak more than one language. The same protocol can have different versions, in the same way French has a northern, southern, and many other dialects. People speaking different dialects may be able to understand each other completely or partially, and it is the same story with protocols that may be completely or partially compatible with each other. Often, there is a **backward** compatibility. For example, a system speaking *ABC Protocol Version 2* can understand another system speaking *ABC Protocol Version 1*. In this case, their communication is limited to what a less capable system (Version 1) understands. This is similar to speaking to a five-year-old child—you can't discuss advanced math topics even if both of you are speaking the same language.

Another important thing to understand is the difference between call control and voice flow in IP telephony. Remember: by call control, we mean the way we control establishing calls and operations such as transfer, hold, picking up calls, and disconnecting them. In the world of legacy telephony, both call control *and* voice flow have to pass through the PABX unit physically. Look at any of the pictures illustrating PABX, and you'll realize there is no other way for voice transmissions to pass from one phone to another except by going through the PABX box.

In the case of IP telephony, the situation is different. Call control is still managed by the IP PABX (i.e., a phone has to speak to IP PABX to tell it that it needs, for example, to establish a call with extension 1014), but once the call is established, the voice flows from one phone to another through the network switch but *not through the IP PABX*.

The last important point is that you'll need a few protocols to make two phones and IP PABX talk to each other. First, you need a signaling protocol. This is the language the phone uses to tell the IP PABX that it needs to dial extension 1234, for example. Examples of signaling protocols include H.323 and SIP, with the latter being the most popular world-standard nowadays. Second, you need a protocol to convert speech to network traffic and send it to another phone (which another phone needs to understand to convert it back to real speech that a human can hear and understand). These protocols are often referred to as **codecs**. There are myriad codecs out there, such as G.711, G.723, G.729, iLBC, GSM, and so on—we could easily fill this entire page with their names if wanted to, but understanding the concept is more important so we won't).

That is enough for today. Be sure you understand and remember all of the concepts in this chapter before moving on to the next one tomorrow. The quiz will help you check your understanding of the topics. Going through the pictures and drawing them yourself as you go through the main concepts will help you understand the material even better.

See you tomorrow in the next chapter, where we start building proper contact center systems and discover the cool functionality they provide.

Summary and Quiz

In this chapter, you learned about the evolution of telephony from the very basic to complex IP telephony systems. You learned about components of IP telephony, their functions, and the advantages IP telephony can bring when compared to traditional telephony. You also learned about some basic functions PABX systems provide. In the next chapter, we discuss how contact center systems are built on the foundation of IP telephony.

Please answer these self-check questions to test your knowledge of the topics covered in this chapter:

1. Which one of the following is not a component of IP telephony setups?
    a. Switchboard
    b. Media gateway
    c. IP PABX
    d. IP Telephone
    e. All of the above are components of the IP Telephony setup.

2. What is the function of CO lines?
    a. To connect office telephony systems to emergency services such as 911.
    b. To connect office telephony systems to telecom companies.
    c. To connect IP telephony systems to the Internet for cheap international calling.
    d. To connect IP telephony systems to traditional telephony systems.

3. Which three (3) of the following statements are correct?
    a. IP Phones connect to networks just like computers.
    b. IP Phones connect directly to IP PABX via cables.
    c. Communication between IP Phones must pass through the IP PABX.
    d. Communication with the outside world must pass through the media gateway.
    e. IP PABX knows all extensions; IP addresses of the phones.

f. IP PABX knows only phone extension numbers; network switches know the IP addresses of the phones.

4. Which of the following is not an advantage provided by IP telephony?
   a. Better security of voice communication.
   b. Less cabling is required as IP phones connect to the same network as computers.
   c. IP telephony does not require any equipment except for IP PABX and IP Phones.
   d. It is much easier to move employees from one workplace to another in an IP telephony environment.
   e. All answers above represent valid advantages of IP telephony over traditional telephony.

5. You have three (3) employees with extensions 1001, 1002, 1003. You want these employees to receive random incoming calls. Which feature should you ask your IP PABX administrator to configure?
   a. Auto-attendant
   b. Virtual number
   c. Randomizer
   d. Hunt group

6. Whenever someone calls your company's number, you would like the caller to hear a welcome message. Which PABX/IP PABX feature can help you in this situation?
   a. Voice mail
   b. Auto-attendant
   c. Virtual secretary
   d. Media gateway
   e. Switchboard operator

7. Your colleague instructs you to dial 9 before the phone number in case you need to make an outside call. What is the function of the 9?
   a. It is a password to activate the phone line for outside dialing.
   b. It is an outside line prefix.

c. It is a code turning on the recording system to record the call.

d. It is the expected duration of the call you are about to have.

8. Your friend, Mike, has an internal extension number of 1099. If you call him from outside of the company, you can reach his phone directly by dialing the number 2341099 without hearing the auto-attendant and without the need to dial an extension. What is the name of the feature Mike has enabled?

   a. Virtual transfer

   b. DID

   c. DTMF

   d. CO line forwarding

9. Call transfer, forwarding, conference, hold, and waiting are:

   a. Functions found only in IP PABX.

   b. Functions found only in PABX.

   c. Functions commonly referred to as "call control."

   d. Functions requiring pulse dialing.

   e. Functions of hunt groups.

10. Which of the following statements is false?

   a. IP PABX performs the call control function.

   b. Network switches should have the quality of service configured.

   c. IP phones usually provide encryption.

   d. Signaling protocol is required for phones to communicate to IP PABX.

   e. All of the above statements are correct.

Day 2: How the Call Centers Work—A Bird's Eye View

Before going into details and drawing diagrams (but I promise, we will), let's take a look at the call center the way a typical consumer does. In this chapter, we focus on the **call center**, a division in a company responsible for communicating with its customers via telephone. Unlike **contact centers**, call centers only deal with phone calls, and this is what we are going to discuss today. We will get to the topic of **contact centers** a bit later.

Assume that one nice sunny day, you decide to order a pizza, and a friend recommends a certain place to you. You decide to order it delivered, so you pick up your mobile phone and dial the delivery number your friend just gave you. A nice voice on the other end says, "Welcome to XYZ Pizza. Please select the language you want us to serve you in. For English, please press one; for Spanish, please press two; for French, please press three."

Don't ask me which country we are in now, but let's imagine you prefer to be served in English (that'd be my preference, too), so you press one.

The voice says, "All our agents are serving other customers. You are number two in the queue and will be connected to an agent in approximately 20 seconds." While you are waiting, a promo message plays, informing that you can get a medium-sized pizza free if you order a large one, and this sounds like a good deal to you.

After waiting for 18 seconds, you are connected to an agent. You tell your order to the XYZ Pizza employee, who takes your delivery address and confirms your mobile number. When all is done, she promises that you'll get your lovely, hot pizza (and an extra free one because you claimed the promo offer) in 40 to 50 minutes, delivered to your doorstep, and she transfers you back to the "machine" which asks you to press a digit from one to five to rate your experience while ordering from XYZ Pizza, where five is the best. You give it a five since the experience was fantastic.

Five minutes later, you realize that you forgot to order a salad for your wife (this is just an example—don't do this in real life), so you dial

XYZ Pizza again. You notice that the greeting is in English (you don't have to choose language again), and a voice says, "Your order is processing now and should be delivered to you in approximately 29 minutes. If you would like to make a change to your current order, please press one. For all other inquiries, please press two." You press one and are connected to an agent almost immediately, add a green salad to the order for your wife, and hang up.

Later that day (after enjoying your pizza and salad, which were both delivered on time), you decide to call XYZ Pizza to give feedback about their service. When you call, you hear the greeting in English, asking you to press one to place a new order or two to provide feedback about the order you received earlier in the day. You choose two, and your call is redirected to a customer survey, asking you a set of questions about your experience. At the end of the survey, the system offers to connect you to an agent if you want to provide extra feedback, which you choose to do. When the agent picks up and thanks you for the survey you have just completed and the excellent score you gave, you tell her that next time you'll enjoy it even more if they provide an extra sachet of black pepper with the salad since your wife likes it spicy. The agent records the feedback, thanks you, and you hang up.

A few days, later you decide to order the same pizza and salad again. When your call is connected, the system greets you in English, and when you reach the agent and tell her your order, she asks if you would like an extra sachet of black pepper added with your salad, and you say that you would. She confirms that your delivery address is the same as the last time, finish the call, and hang up. You get your order 43 minutes later and recommend XYZ Pizza to colleagues at the office the next day when they decide to order a pizza for lunch.

This illustrates an example of a perfectly working call center and how it helps XYZ Pizza make its customers happy and even attract new customers. It is now the time to see the modules/components/systems (whatever you may call them) that make this all happen and how these bits and pieces work together to achieve the great **customer experience** you rated with five stars.

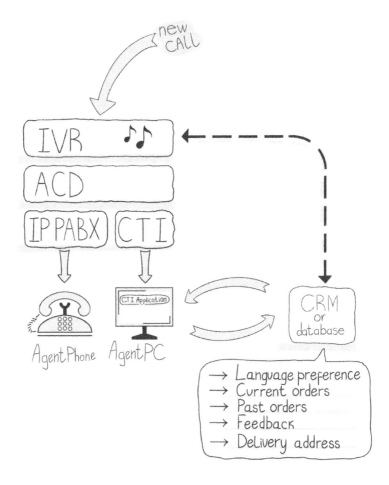

The first system (the one that greets you) is typically either an **auto-attendant** system or an **interactive voice response (IVR).** These systems play similar roles—playing voice announcements to callers— but there is a significant difference between them, which makes IVR more common in contact centers. Auto-attendants can have simple menus to transfer your call to the desired destination or play you a pre-recorded message based on the keys you press on your phone's keypad. This is pretty much all that auto-attendants can do. The algorithm auto-attendants use to route calls never change on the fly, except maybe in obvious scenarios in which you might greet people differently at night or on the weekend than you do in the middle of the workday.

IVR is a more sophisticated creature. Besides advanced *call flow* scenarios, IVR can also "talk" to other systems, either putting information into the system or requesting information from it. If you go back to our example above, you will notice that when you called XYZ Pizza the second time, the IVR did not ask about your preferred language. This is because the IVR saved your preference in the database when you called for the first time. When you made your second call (and all subsequent calls), the IVR detected your mobile number and went back to the database to check if there was a record about a preferred language associated with the number. Based on your phone number, IVR is able to retrieve a lot of data from the database. In this case, it retrieved your preferred language and if there was an order pending or an order recently delivered (in case you are offered the option to provide feedback).

Very often, the IVR plays another important function: the **queue**. This is where calls are connected while they are waiting to be served by an agent. While a customer is in the queue, the IVR can play a promo message or standard, pre-recorded message. It is a good idea to customize and play different messages to customers, depending on what you know about them. For example, a bank may choose to play a credit card promo message for customers in the queue with a certain amount in their accounts.

The next system is an **automatic call distributor (ACD)**, which is like the brain of the call center system. The ACD does not accept any calls itself, like the IVR or agent phone, but it intelligently decides what to do and tells other systems what they should do. ACDs also track which agents are available by agent **skills**, the ability of a given agent to provide a certain service. A typical example of a skill is the ability to speak a certain language or provide a certain service in that language. For example, assume the following matrix of skills is available:

- EN_Orders for taking orders in English;
- FR_Orders for taking orders in French;
- EN_Complaints for responding to customer complaints in English; and
- FR_Complaints for responding to customer complaints in French.

The ACD constantly monitors the agents to know which agents are available, how many there are, and can even predict when the next agent with a given skill will become available with a certain level of accuracy. When you choose the option in the IVR to talk to an agent, it will pass the request to the ACD needed to transfer the call to an agent, and the ACD checks for an available agent. For example, say we need an agent with the EN_Orders skill. If the agent is available, the ACD will respond to IVR with the agent's extension so the IVR can transfer the call to that extension. If the agent is not available, the ACD responds to the IVR to put the call into the queue. Once an agent becomes available, the ACD comes back to IVR, asking it to retrieve the call from the queue and transfer it to the agent. We should mention that the same agent may have more than one skill assigned to him. For example, if a person speaks more than one language or is able to do different kinds of tasks effectively. Most of the ACDs I know support multi-skilled agents.

ACDs also monitor **agent status**, which is what an agent is doing at any given time. Examples of agent statuses are:

- **Ready**: the agent is ready to receive the call immediately;
- **Not ready**: the agent is not ready to receive the call for whatever reason;
- **Talking**: the agent is currently talking with another customer;
- **Logged out**: the agent is logged out of the system, most probably because this is not his/her shift; and
- **Wrap-up**: the agent is finishing some work after a customer call he/she just had. The wrap-up can be automatic (the ACD automatically puts an agent in a wrap-up state after each call for X seconds) or manual, when an agent has chosen to go to a wrap-up state when required.

After reading the last few paragraphs, you might ask if we can still put the equals (=) sign between the terms **queue** and **skill**. The answer to this is YES. For example, since XYZ Pizza needs to have people taking new orders in English, they configure this skill in the ACD, assign it to a few agents, and create a queue. If at the same time you request to speak with an agent and there is at least one *ready* agent with the required skill, then your call will be transferred immediately, but if there are no agents available, the ACD will ask the IVR to put you into the queue. Because the ACD continuously monitors how calls are received and handled, it can statistically predict the next available agent in 20 seconds, which was a rather accurate prediction as you were transferred 18 seconds later.

Let's use the diagram below to illustrate how skills and queues work together and how calls are distributed to agents.

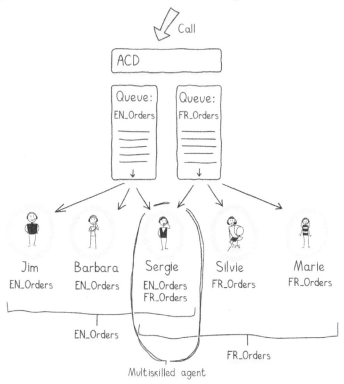

In the above diagram, we have an ACD with two skills configured: EN_Order and FR_Order. There is a total of five agents—Serge, one of them, is a **multiskilled** agent, having more than one skill. The default distribution mechanism for most ACD systems is the **longest available**. In other words, if Serge is currently on a call, Barbara has been ready for two minutes, and Jim has just finished a call 30 seconds ago, then Barbara will be the one to receive the new call when it arrives at the EN_Order.

There is also a way to configure agents with different skill **proficiency** levels and configure the system to use the agent with the highest proficiency first. For example, if Jim has a proficiency level of eight assigned to him, Barbara has a six, and both of them are ready, then Jim will get the call if the ACD is configured to route to the most proficient agent. If both agents are ready and have an equal proficiency but one has been available for a longer time, then the ACD will normally send the call to the one who has been available longer.

We have just discussed how calls are distributed to agents in the same skills group, but it is a good idea to see what happens if one agent is assigned to two queues, which is Serge's situation, as she speaks both English and French. Let's imagine a situation in which all agents—Jim, Barbara, Silvie, and Marie—except for Serge, are busy. There are two calls in two queues—one in EN_Order and one in FR_Order—but we only have one agent available that can take either of them. Which of the two calls is assigned to Serge depends on a few factors.

If queues have equal **priority** (i.e., they are equally important to us from a business perspective), then the call that has been in the queue longest will be transferred first. For example, if one customer is waiting 25 seconds and the other has been waiting for 13 seconds (each is in his own queue, but time is the absolute measurement here), then the customer waiting 25 seconds will be first to get transferred.

You may also have a situation in which queues have different priority levels. For example, if FR_Order has a higher priority than EN_Order, then the call in the FR_Order queue will be transferred to Serge first. This approach is very useful for VIP treatment—creating two queues/skills for the given agent's capabilities, for example, answering credit card-related questions at the bank. In this case, you would create two skills: CreditCardNormal and CreditCardVIP and two queues with the same names. You would put a higher priority on the CreditCardVIP queue, so if there is at least one call in that queue, it will be transferred to agents first. When the CreditCardVIP queue is empty, calls from the CreditCardNormal queue are transferred in first-come-first-served order. You may see the same approach at airline check-in stands in some airports—when there are no business-class passengers, the business-class stand calls normal passengers to check in there, but if a passenger traveling in business comes, he will be served at the business-class desk as a priority.

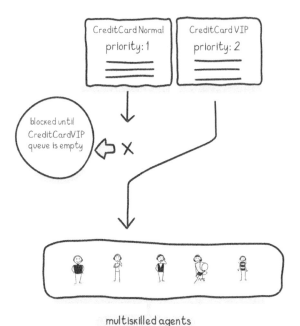

CreditCard Normal
priority: 1

CreditCard VIP
priority: 2

blocked until
CreditCardVIP
queue is empty

multiskilled agents

This approach is reasonable as long as you don't have too many VIP calls, or calls in the low priority queue may never get the necessary treatment, and the **service level (SL)** will suffer. Service level is an important performance indicator in the call center, and despite the fact that we have a separate chapter for **key performance indicators (KPIs)**, it is worth explaining the term service level here, which is a basic way to see how well (or poor) the call center performs at any given time.

Service level is applied and measured on the queue level (so different queues will have different service levels). SL is a measurement of the percentage of calls answered by agents within X seconds. For example, we may set the service level at 20 seconds, only to find that 92% of the calls are answered by agents in less than 20 seconds (time starts counting from the moment the customer chooses the option "talk to agent"). Contact centers often define the **service level target** as a two-digit number representing the service level they wish to achieve. An example might be 80/15, which means that 80% of the calls must be transferred to an agent in less than 15 seconds. The **actual** service level may be 83/15, meaning that the call center is doing well—or 60/15— meaning that it is doing not as well, and people are left to wait in the queue for a long time.

In the last example, we may find that the CreditCardVIP queue has a good actual service level while the CreditCardNormal one has a lower number, indicating that it is suffering. We will discuss how to correct this later in this book, but for now, let's go back to the core components of the call center—the main topic of today's chapter.

When an agent with the right skills is in a *ready* state, the call is transferred to that agent. To accomplish this, the ACD instructs the IVR to do the transfer. Because the call is transferred to the agent's phone, the IVR sends the request to the IP PABX, and the IP PABX completes the transfer. The ACD controls the process, simultaneously speaking to one more important component, called **computer telephony integration (CTI).** The role of the CTI is to display the information about the caller on the agent's PC screen at the same time as the phone rings. CTI applications running on the agent's PC can also bring information in from an external database to the agent's screen.

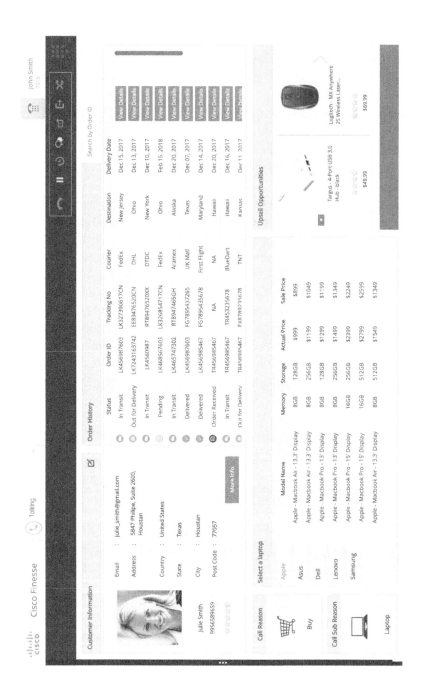

Novelvox Unified Agent Desktop application showing caller details running inside Cisco Finesse (registered trademark of Cisco Systems) CTI. © Novelvox. All rights reserved worldwide. Used with permission.

For example, when you make your second order, the agent can already see all your details: the address to where the order should be delivered, your name, past orders, and even that you prefer an extra sachet of black pepper with your salad. This helps the agent finalize your order quicker.

A CTI application may run as a **thick client** (i.e., like the software you need to install on an agent's PC) or **web-based** (or **thin client**), which runs inside an Internet browser. The example above shows the Finesse CTI application from Cisco Systems, running inside Mozilla Firefox. The way the CTI runs does not make a lot of difference, but the contact center industry generally moves toward web-based applications. One simple reason (but it is not the only reason) for this trend is that web-based applications are easier to maintain. When apps are in need of updates, you only have to do it once on the server and ask your agents to refresh their browsers (the same as with a web page). With a thick client, the story becomes a bit more complex when you have a large number of agents and agent PCs to maintain.

We should mention the **customer relationship management (CRM)** system. Strictly speaking, this system is not part of a call center, but it plays an important role in making it effective. CRMs store customer information, such as names, phone numbers, addresses, and details of all that happens between the company and the customers, such as orders, complaints, preferences, and the like, regardless of whether it happens in the call center or any other way. This is why CRMs are *not* and *should not* be a part of a contact center but central to the entire company, and the contact center must have direct access and be able to work with it. If you order a pizza over the counter, the counter staff should have access to the information in the CRM. If you order a pizza through a mobile application or web site, the situation should be the same; XYZ Pizza stores all of your details in one centralized location to serve you better, regardless of how you place your order. This is why the CRM must be centralized. It is also the key to providing consistent customer experiences.

The picture below shows a CTI interface with the CRM integrated in it. When a customer's call is transferred to the agent, the CTI automatically opens the CRM and displays the customer record for the given customer. This saves a lot of time for both customers and agents and should be considered best practice for any call center.

r applications besides CRMs can be opened as a part of the
lication. In the example below, the CRM and mapping
ions are both opened inside the CTI to show information about
the current calling customer:

With so much being discussed about the agent's workplace, it is
now time to stipulate that not all agents are equal and some of them
have a few more capabilities than others in their CTI; these agents are
normally referred to as **team leaders** (TL) or **supervisors.** "Team
leader" is more common in the industry, so we will use it in this book.
Also, depending on how the hierarchy of the contact center is defined,
it may have both TLs and supervisors, meaning it has different levels of
seniority.

The team leader is a more experienced agent with slightly elevated
privileges in the system. Normally, there is one team leader for
approximately every eight to 12 agents. The team leader usually has
the skill to answer customer calls (like normal agents) but he will
typically answer calls at busy times or when a customer asks for an
escalation (saying something like, "Can I speak to your manager?"
when something goes really wrong or a customer is not happy with the

agent's answer, and he wants to confirm it). Other times, the TL monitors how the agents in his team perform. Depending on the manufacturer of the contact center system, the TL may have a few additional functions on his CTI interface, such as:

- **silent monitoring**, when he can listen to live conversations between agents and customers;
- **barge-in**, when he can enter a live call and participate in what is happening between agents and customers;
- **intercept call**, when he can take calls from the agents;
- **view and control** the status of agents in his team, manually changing an agent's status "not ready," for example;
- **view and control** the status of queues according to the skills on his team;
- **chat or send** instant text messages to an agent's CTI interface; and
- **whisper coaching**, when an agent is on a call with a customer and the TL can speak on the call, but only the agent (and not the customer) will hear.

Team leaders usually report to supervisors or directly to contact center managers, depending on how large the call center is and how complicated the company structure is. Team leaders don't deal with customer calls (at least 99% of the time) and use systems to manage and monitor contact center operations. We touch a few of those here.

- **Reporting database** – is where the IVR and ACD log information about everything that happens in them, including calls received, time to transfer calls to agents, options customers choose in the IVR, number of agents available, calls received, the service level at any point in time, which agent changed states and why, and so on. Reporting databases literally have hundreds of metrics to measure, which contact center managers view via reports.
- **Historical report** – shows metrics over a period of time, for example, the average service level in 30-minute intervals over the past week for the En_Order queue.
- **Real-time report/dashboard** – shows metrics as they are in real-time (or almost real-time). For example, the number of agents logged on and ready in a given team.

- **Wallboard**—a report available on a large screen hanging on a wall. It can show the same values as a dashboard but is useful for displaying information for everyone in the area to see, for example, the real-time information on service levels for all queues. The difference between a dashboard and a wallboard is that a dashboard is more personal, and it shows important employee statistics, which may or may not be confidential running on a PC screen. A wallboard, as the name suggests, is real-time information displayed on the wall for everyone to see.
- **Workforce management (WFM)**—a system helping management to create agents' schedules to see how they adhere to these schedules. We will cover more on WFM in the following chapter, Workforce Optimization (WFO).
- **Voice Recorder**—a system recording all calls between customers and agents. There are myriad reasons to record calls in a call center. Again, this will be discussed in further detail in the WFO chapter of this book. Note that voice recorders typically take data from the ACD, IP PABX, and the phone itself. Besides recording voice conversations, it records additional data, such as the phone numbers of callers and agents, which agent answered which calls, from which queue the call came, and so on. In the WFO chapter, we discuss how this data can be useful.

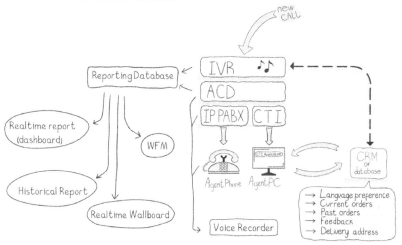

To perform adequately, team leaders will have slightly advanced versions of the CTI application used by agents, with the necessary controls to perform the functions above. The picture below shows an example screen of a team-leader agent using the CTI application, showing the state of agents on his team:

Cisco Finesse

Team1 — Talking — Agent State

John Smith

Skill Group Statistics Display | CSQ Queue | Precision Queue | Follow Up | Scripting Summary | Add Field

Search by Agent Name

AGENT NAME	AGENT STATE	DURATION	DIRECTION
Cheryl Elvis	Talking	01:30	IN
Cecile Young	Ready	00:45	NA
Elvis Dave	Not Ready	01:01	NA
Cherina Essy	Talking	02:05	OUT
Elvis	Work Ready	02:00	NA
Young	Login	02:45	NA
Rhinda Greer	Talking	01:45	IN
Michele Hunter	Talking	02:45	IN
Todd Jenks	Talking	01:33	OUT
Elvis Dave	Logout	01:45	NA
Todd Jenks	Logout	00:45	NA
Elvis Dave	Logout	02:11	NA
Terrence Reed	Logout	01:44	NA
Paige Hunter	Talking	03:00	IN
Jonathan schofield	Logout	02:15	NA
Dexter Young	Talking	01:45	OUT

Supervisor Actions: Signout | Ready | Start Monitoring | Start Barge In

Search by Skill Name

Skill Name	Logged In	Available	Unavailable	Total Calls	Oldest Contact	Handled	Abandoned
Credit Card	56	50	6	30	00:02:34	23	34
Net Banking	45	40	5	45	00:01:34	45	32
Debit Card	48	44	4	325	00:06:23	323	23
Loan	62	56	6	35	00:02:34	34	45

Team Statistics Information

Search by Agent Name

Agent name	Login Time	Calls Presented	Calls Handled	Total Login Duration	Max Talking	Avg Talking	Total Talking
Cheryl Elvis	00:01:09	23	23	04:55:26	00:02:34	00:01:09	00:02:34
Cecile Young	00:01:09	2	2	04:55:26	00:02:34	00:01:09	00:02:34
Elvis Dave	00:01:09	23	23	04:55:26	00:02:34	00:01:09	00:02:34
Cherina Essy	00:01:09	2	2	04:55:26	00:02:34	00:01:09	00:02:34
Elvis	00:01:09	23	23	04:55:26	00:02:34	00:01:09	00:02:34
Young	00:01:09	2	2	04:55:26	00:02:34	00:01:09	00:02:34
Rhinda Greer	00:01:09	2	2	04:55:26	00:02:34	00:01:09	00:02:34
Michele Hunter	00:01:09	2	2	04:55:26	00:02:34	00:01:09	00:02:34
Todd Jenks	00:01:09	2	2	04:55:26	00:02:34	00:01:09	00:02:34
Elvis Dave	00:01:09	2	2	04:55:26	00:02:34	00:01:09	00:02:34

Chat with Agent | Recording

Novelex Supervisor application based on Cisco Finesse CTI Framework (registered trademark of Cisco Systems) © Novelvox. All rights reserved worldwide. Used with permission

As you may see from this example, the team leader is concerned about one of the agents, Dianne Ignacio, who has spent more than three hours in a "not ready" state.

Managing agent performance is one of the cornerstones to building an efficient contact center. In chapters one and two, we discussed the technology in contact center work, but as we said earlier, a contact center is not only about technology—it is also about people and processes. In our next chapter, we discuss how to optimize and manage the contact center's primary workforce: the agents.

Before we go to the next chapter, I suggest doing a small study using Google to try to find information about the cost structure of a contact center to figure out the most expensive part of contact center operation. Take a look at:

- the cost of the technology (ACD and IVR software, cabling, servers, and so on);
- the cost of telecommunication (what we pay to the telco for our phone lines);
- the cost of people (salaries, training fees, and other expenses associated with employees); and
- the cost of facilities (such as building maintenance, utility charges, furniture, and so on).

You'll see that one of the items on the above list costs the lion's share more when compared to the rest of them combined. This explains why the next chapter—on Workforce Optimization (WFO)—is so important to read and understand.

Summary and Quiz

In this chapter, you have read about the experiences customers have with call centers. You have also learned about components comprising call centers and their functions. You learned the basic principles of call distribution, tools used in call centers, and how agents and supervisors use these tools to perform their day to day tasks. In the next chapter, we discuss how to optimize this efficiently working mechanism even further.

Please answer these self-check questions to test your knowledge of the topics covered in this chapter:

1. When you dial a call center for your bank, you hear a message saying, "Thank you for calling ABC Bank. To continue in English, please press one." The component responsible for playing this message is called:
   a. Interactive Voice Response (IVR)
   b. Computer Telephony Integration (CTI)
   c. IP Phone
   d. Voice recording system
   e. Voice playback system

2. Which application works on the agent's PC and is synchronized with the IP PABX and the agent's phone?
   a. Automatic Call Distribution (ACD)
   b. Computer Telephony Integration (CTI)
   c. CRM system
   d. Network switch

3. What are the functions of the ACD (choose three (3) answers)?
   a. Perform skill-based routing based on an agent's skills.
   b. Play messages to callers.
   c. Create agents' schedules.
   d. Monitor agent states (ready, not ready, etc.).
   e. Log information into a reporting database.

4. A service level of 80/20 means that
   a. 80% of the calls must be answered by agents, and 20 are transferred to team leaders.
   b. 80% of calls must be answered, and the remaining 20 can be abandoned.
   c. Calls must be answered by agents within 20 seconds, and agents have 80 seconds to handle each call.
   d. 80 seconds is the maximum time calls will spend in the IVR system, and 20 seconds is the maximum time calls can spend in the queue.
   e. 80% of the calls must be answered within 20 seconds.

5. A call center processes calls for normal and VIP customers. You want to make sure that the VIP calls will be processed with a higher

priority, so you create two skills and two queues for normal and VIP calls. All agents in the call center are multi-skilled and capable of processing calls from both normal and VIP customers. What is the right way to configure the call center?

    a. Configure the VIP queue with a higher service level target than the normal queue.

    b. Configure the VIP queue with a higher priority than the normal queue.

    c. Configure the actual service level on the VIP queue to be higher than the actual service level of the normal queue.

    d. Configure the actual service level on the VIP queue to be higher than the target service level.

    e. Configure the target service level on the VIP queue to be higher than the actual service level.

6. You are managing a large contact center. The call-taking procedure includes filling out a small form after each call is completed to leave information about the call type and outcome. The form is integrated into the agents' CTI screen. Some of the team leaders complain that after a call is finished, the agents don't have enough time to fill out the required form. What is the best way to solve the issue?

    a. Ask the contact center support team to reduce the target service level by five seconds.

    b. Ask the contact center support team to automatically extend agents' "talking" state for an extra five seconds after a call is finished.

    c. Ask the contact center support team to increase the wrap-up time by five seconds.

    d. Ask the contact center support team to introduce the delay in the ACD system to pause for five seconds before transferring the call to the agent.

7. This component is responsible for displaying information about callers on the agents' screens at the same time as the agents' phones are ringing. What is the name of this component?

    a. CTI with integrated CRM pop-up

    b. CTI with historical report system pop-up

c. CTI with real-time dashboard pop-up
d. IPPABX
e. Supervisor call intercept feature

8. The team leader would like to monitor some of her team members' live customer calls without them knowing it. What is the feature the team leader should use?
   a. Barge-in
   b. Intercept call
   c. Dashboards
   d. Silent monitoring
   e. Direct monitoring

9. What is the main reason CRMs should not be a part of a call center but instead must be tightly integrated with it?
   a. CRMs are produced by different vendors than contact center platforms.
   b. CRMs should accumulate information about customers regardless of their channel of communication (call center, branch, website, mobile app, etc.).
   c. CRMs are difficult to integrate with contact center solutions.
   d. CRMs don't have IP PABX functionality.
   e. All of the above answers are incorrect. In fact, CRMs must be a built-in part of a call center system.

10. What information can be shown to an agent when the call is received if the contact center has an integrated CTI with business applications? (choose all that apply):
    a. Customer's phone number
    b. Customer's location on the map
    c. Customer's history of previous calls
    d. Customer's name and contact information from the CRM

Day3: Workforce Optimization (WFO)

The format of this book leaves me only one chapter to discuss the topic of workforce optimization (WFO), which means that we will only slightly scratch the surface. I will explain the core concepts and elements of WFO here and suggest the reader do further research once the basics have been "digested" and understood.

Workforce optimization is such a broad topic (and industry) that many companies specialize solely in WFO tools—there are books, seminars, and scientific research regarding WFO, and they all make sense. If you completed yesterday's research, you have most probably figured out that the agents (generally speaking, the workforce, which includes the agents, team leaders, managers, and other people working in the contact center) is, indeed, the most expensive part of contact center operations, accounting for 60-70% of the overall expenditure. By **workforce costs**, we mean direct costs—such as people's salaries—and indirect costs—for hiring, training, paying taxes, social insurance, and the like. Normally, the cost of contact center software licenses (which every agent consumes) is considered a part of the **technology cost**, and the cost of furniture, cooling, and premises are considered a part of the **facilities cost**, but this depends largely on the calculation method. The most important outcome is that a contact center operation may save a lot of money if its management finds a way either to process more calls using the same workforce or process the same amount of calls with a smaller workforce. This simply means it costs less money to process each call.

With the above math, you can easily justify an increase in technology spending if it reduces your agents' costs by the same percentage. For example, assume we have a contact center with an annual budget of USD $1,000.000. As you can see from the formula above, USD $650,000 is the approximate workforce cost. This leaves $150,000 for telecom and facilities costs, with $200,000 remaining as a technology cost. If we increase technology costs by 10%, this brings us to $220,000 (an extra $20,000), but it will potentially reduce the workforce cost by 10% ($65,000), so we will gain $45,000 annually.

Fairly speaking, I may be criticized for this example, and rightly so, because it looks at the contact center as a **cost center**. Cost centers happen when companies establish contact centers simply because they must have one. This may be because supporting customers through other channels (such as branches) is even more expensive. Sometimes it is because the end-users expect your business to have some kind of contact center, for example, if you are a bank or a telecom company. In reality, contact centers (when properly designed and managed) are no longer simply cost centers anymore—they are also **profit centers**, which means that they generate money for the company (or directly help the company to generate money). One good example of a contact center as a profit generator is the contact center of an airline. When properly built, such a contact center may answer customer inquiries (cost) and sell tickets or class upgrades (profit) at the same time. A bank's contact center can facilitate customers paying bills via phone banking as well as asking customers to apply for new credit cards, both of which generate a profit for the bank. You can order a pizza online, and a follow-up call from the call center will inform you that you can get an extra salad at half-price, and it is very likely that you will upgrade your order in the same call. In the above examples—airlines, banks, food delivery—contact centers directly help companies generate more money.

The statements above should be corrected in the following way: workforce optimization technology allows us to:

- serve more calls (or contacts)
- with less workforce
- with better quality and consistency of service provided
- with the net result being a higher profit and less cost (i.e., higher net profit).

To go even further, because contact centers nowadays are the largest touchpoints between businesses and their customers, they became the cornerstones for building something called "customer experience," or CX. Although we will talk more about CX later in this book, for now, it is important to understand that a contact center operating with a consistent quality helps a company with profitability and good customer experience. To operate the contact center with consistent efficiency and quality, we need a solution called Workforce Optimization or WFO.

Let's begin with the basics—what is WFO?

Workforce Optimization (WFO) is technology (a framework of tools, applications, and technologies), that help contact centers to *optimize* agents count and facilitate the highest level of service and quality of customer experience while keeping the costs reasonable and (where applicable) achieving high profits.

There are many large vendors in the market, providing WFO solutions. Some offer a subset of tools, while others offer a complete ecosystem that goes far beyond WFO. Each of these vendors has its own way to group its products, and this is why it is difficult to build a structured diagram representing a common industry view. In this book, we are only going through the most important parts of WFO, so I will summarize these in the picture below and explain each of the families in WFO: recording, quality, and workforce management.

WORKFORCE OPTIMIZATION

Recording & monitoring	Quality management	Workforce management
Voice recording	Agent evaluation	Traffic forecasting
Screen recording	Voice of the customer	Agent scheduling
Live monitoring	E-learning	Shift trading
Speech analytics		Real-time management
Desktop analytics		Payroll integration
Motivation and gamification		

Workforce Optimization Products

Recording and Monitoring

Voice Recording is a module responsible for recording voice conversations (or interactions) between agents and customers in the contact center. There are several reasons why voice recordings in the contact center may be important, such as compliance, quality, and security, among others. If you go to a telecom company's branch and ask to upgrade your package, the employee there will probably give you a paper or two to sign. When you make the same request through a call center, you cannot physically sign anything, so the recorded call is the only evidence your mobile operator has to prove that you requested the upgrade (in case you complain later after receiving a higher bill at the end of the month). Having such evidence is usually a government requirement, and all banks, telecom companies, and other service providers (through call centers) must have a recording system in place to comply with the regulations. Different countries require recorded calls to be stored for months, years, and sometimes even decades. Because all companies must comply with government regulations, these types of recordings are usually called **compliance** recordings.

Another reason for recording calls is to use them for further analysis of the interaction itself, in a process called **quality evaluation.** We will talk about this later in the chapter while discussing the **quality management** module of WFOs. There is one more use for recording voice files, which should be mentioned here: security. This is extremely important in the banking and financial industries. If hackers call into call centers to try to gain illegal access to bank accounts (using their *social engineering* skills), their voices will be recorded, which can later be used for investigating the case. We can also mark known fraudsters' voices in the system and compare incoming calls to known fraudsters in the database and send an immediate alarm to an agent if the system detects a match. This use can be extended with **voice authentication**. If person A calls to the call center and passes authentication through IVR using his/her PIN, ID number, and so on and later talks to an agent, his/her voice is stored, so when the same person calls back, the voice can be compared to the stored sample as a security authentication mechanism. In the same way you use your fingerprints to access your office or your PC, the call center can use the stored voices as "voiceprints" to authenticate customers and reduce the need for remembering passwords and PINs.

We have answered the question as to **why** we record conversations, but we still have not discussed **how** we do it. If we go back to the diagram of IP telephony setup, you will see that calls between agents and customers pass through three devices: the media gateway, the switch, and the phone. Any of those devices can **fork** the **media stream**; in other words, create copies of both the agents' and customers' voices and send them to the recording system. The recording system normally talks to the PABX unit to get other relevant information about the call, such as the phone number of the customer. It is also beneficial to integrate recording systems directly into contact centers to store call-related data together with recorded calls. Examples of this data (called call metadata) typically include customer ID, the type of call, the skill of the agent, which services the customer used in the IVR before the call was transferred, and many others. This data can be useful for searching calls in the recorder database, as one can filter calls by date, time, duration, type of service, customer segment, and so on. This kind of filtering is even more important in **quality evaluation**. If, for example, we want to evaluate how quickly agents manage to block credit cards on customer requests, we can filter recorded calls by this type of service and quickly select and listen to all calls—the more data associated with the call, the better.

Search and play interface of the ZOOM voice recording system showing call data attached. © ZOOM International, all rights reserved worldwide. Used with permission

The last topic to discuss is where the calls are stored. Without going deep into detail, there are two parts to store: the actual calls (the voice) and something called **metadata**, information about the calls, such as caller numbers, chosen skills, customer IDs, and any other attached information. The call metadata in the database and call files are typically stored separately on large hard drive arrays (sometimes as large as a house refrigerator). Because the voice files are relatively large in size (there are literally millions of them, and each has the size of a few megabytes), **compression** is often applied to reduce the file size and the space they take. Because voice files may also contain sensitive information (especially in the financial industry), **encryption** is almost always applied to these files by default. Because even the biggest storage device is not infinite in size, a **retention policy** is normally defined in the recording system, describing how long the calls should be stored before they are deleted to free up hard drive space. **Selective retention policies** can be applied to various types of calls. For example, one may configure the system to store calls relevant to high-value credit card transaction confirmations for seven years, while all other calls are only stored for three years and then deleted.

Screen recording—saving images of the agent's screen during a call—often works in parallel with call recording. Screen recording is used because contact centers often like to have a record of what agents are doing on their PCs while taking customer calls. These recordings can be used to evaluate agent performance (are the agents really doing what they supposed to and how they are supposed to?). They can evaluate application performance (for example, when looking through the video, the contact center supervisor might discover that the order-placing application freezes for five to ten seconds every time an agent submits an order for a customer, making them wait longer on the line and the call handling extended by this time). They can also assess for lack of automation (for example, if every time an agent has to find the value of an outstanding bill, he has to copy the phone number from one application and paste it to another, losing three to five seconds per call). Screen recordings can reveal a great deal of useful information if properly organized and monitored by the contact center manager. Obviously, watching *every* call is literally impossible; therefore, strategies are often put into place, dictating which types of calls should be analyzed and how often. Videos are played in sync with the agents' and customers' voices, so one may have a complete understanding of what happened during the calls.

With screen recording providing a great deal of potentially useful information, there are few technical concerns associated with it, which we should mention:

- **Size**—video recordings of calls consume eight to ten times more storage space than voice-only recordings. There are couple of solutions to this dilemma: either the contact center will choose to record only five to ten percent of the calls with video (because this will give more than enough call material to evaluate), or the period for retaining video recordings can be made significantly smaller. For example, we may want to store all recorded voice calls for seven years while saving the associated screen recordings for six months only. Another way to deal with size is compression, but a high compression can greatly reduce the quality of video files, making them unusable for analysis, so it is typically better to perform fewer screen recordings but with a higher quality so they can be easily analyzed to provide valuable information.

- **Method of recording calls**—usually, an application runs on the agent PC to record screen images and transfer them to the recording server.

- **Synchronization of video with voice**—achieved with advanced algorithms in the recording application able to read information from multiple sources, including the PABX signaling the call established (at the beginning of the recording) and finished (at the end of the recording).

- **Security**—the contact center agents may have a large amount of sensitive information displayed on their PC screens. As already mentioned, all voice recording files are usually encrypted by default. Screen recordings with sensitive information (such as credit card data) should not only be encrypted, but because of the sensitivity of such information, it may fall under government regulations requiring stronger levels of encryption. Another issue here is that even if the screen recordings are encrypted in storage, contact center staff responsible for viewing the files will have access to a lot of data, they were not necessarily meant to access. The solution for this dilemma to mask sensitive fields on-screen recordings in the resulting video file (for example, credit card data fields) or pause screen recordings when an agent accesses screens with sensitive information.

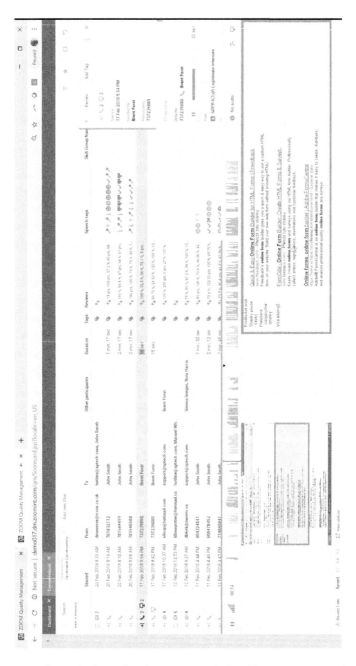

ZOOM WFO--Playing back a call with screen recorded and focusing on a specific screen area where an agent is scrolling through search results. © ZOOM International, all rights reserved worldwide. Used with permission

Live monitoring of agent calls and screens is an added advantage of on-the-go contact center management. It allows contact center supervisors to see and hear what is happening and sometimes make corrections in real-time. For example, a supervisor can chat with an agent and help him/her manage a difficult call with a customer or provide on-the-go coaching instructions to the agent as a part of routine agent training. Sometimes, live monitoring does not require a sophisticated system but a simple device—called a Y-splitter—to split the sound in the agent's headphones.

Y-splitter can be used to connect two headsets to one telephone with one of them used by the team leader to monitor the call

Quality Management

A contact center may have hundreds of users and receive thousands of calls and other interactions—chats and emails, collectively called **digital** interactions, and discussed in the next chapter—every day.

If you call the same contact center a few times, the chance that you will speak to the same agent again is minimal. Regardless, you will probably notice that all agents greet you with the same phrase, often speak to you in the same way, ask you the same questions (such as security questions to verify your identity), and offer additional help at the end of the call. This is because the process of answering calls and making conversation with customers in the contact center is optimized for a maximum ratio of efficiency, and if our contact center is a profit-generating one, maximum profitability.

Let's show an example here. If one wants to work as an agent in a bike shop contact center, it will not be enough for him or her just to be a good bike salesperson (though this is an important prerequisite). Interacting with people over the phone—especially when you are typically making over a hundred calls a day—requires some optimization. You want to greet customers almost automatically (i.e., without thinking about it), you want to remember to offer the new promotion for a free bike check-up for all existing customers, you need to mention that a new model of bike saddle is on limited time discount for all customers, and you also need to tell every customer that your bike shop is arranging a group ride the following Sunday. At the end of each call, you must also remember to ask customers if they have any questions. You must also go through this process approximately a dozen times every hour for the whole working day. If you forget to offer any of the above, your bike shop may lose potential profits. If you spend too much time on a call remembering what to say, your call will last longer, increasing the contact center's operating costs. To optimize the process, agents are trained to answer certain types of calls, what to say, when to say it, what questions to ask, what answers to expect, and how to act on the information they get per the standard procedure for each call in each situation. It is the role of the **quality manager** to control and optimize how agents follow these procedures. Quality managers ensure that agents do what they are supposed to, with the ultimate goal of providing high-quality service for customers in a cost-effective way.

The process follows the usual methodology of plan->implement->evaluate->correct->repeat, which, with some variations, is found in many process optimization guidelines.

Quality management tools help us with steps two, three, and four of the above (almost all of the steps, considering that "repeat" is not really a step).

Call evaluation is usually considered the primary function of quality management systems. This has three main functions:

1. Build evaluation forms, similar to questionnaires—for example, questions might include, "Did the agent introduce himself at the beginning of the call?", "Did the agent remind the customer to keep his personal contact details updated

with the bank?" and so on. Each question has choices and an associated score.

2. Evaluate calls using the evaluation form—quality managers listen to calls and answer questions on the form. Once the process for a given call has finished, a quality score is generated. Obviously, the higher the score, the better.

3. Analyze the results of the evaluations—the methodology of analysis depends on the overall evaluation strategy (i.e., What we want to find).

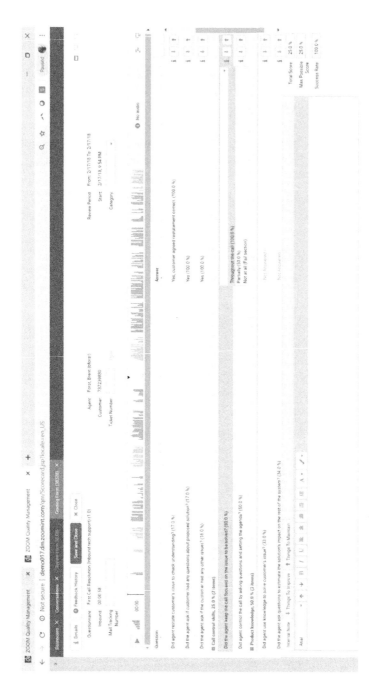

Agent quality evaluation screen with questions--ZOOM WFO solution.   © ZOOM International, all rights reserved worldwide. Used with permission

All three functions are closely related to each other and the overall goal the quality department wants to pursue. If we want to improve the overall contact center quality, calls can be selected at random, and questions will contain only high-level information about greeting the customer, finishing calls, ways of speaking, and the like. If we want to evaluate a certain group of agents, then the selection of calls for evaluation will be from that group only. If we only need to evaluate a certain function—such as account services in a bank call center—then a random selection of calls will be made from customers choosing "account services" before they are transferred to agents. If you remember, we mentioned that recording systems could be intelligent enough to store call metadata (information about calls) together with the recorded voice. In this example, the type of calls is "account services," and you can see how useful and important this feature is in the quality process; now, we have a tool with which to complete selective quality control on a certain function in the call center.

Agent eLearning is a function that helps train agents. There is nothing new in eLearning—it is a set of digital materials, including texts, video, audio, and graphic information the agents have to study. It is also a system with which to assign these materials to agents and control how they work through the courses. The value of having eLearning as a part of the overall WFO package is that it is a part of the overall cycle described above. For example, we can automatically assign certain courses to agents with lower quality scores. After the agents pass the course, they can be re-evaluated to measure the positive impact their training had. Another application eLearning is learning by example. We can find some "good" or "bad" calls processed by experienced agents and use them as material to illustrate what to do and what not to do with real examples. Another value to having eLearning as a part of the overall WFO package is its linkage to the workforce management system (WFM). We will discuss the WFM in a little while, but for now, it is enough to understand that the role of the WFM is to help the contact center to schedule the necessary number of agents with the necessary skills and qualifications to be available in the call center when needed, and modules are directly related to training agents to have the necessary qualifications. Therefore, having eLearning as a part of the WFO (rather than separately) may offer benefits such as those described above as well as others, depending on the way the functions of a certain product are integrated and implemented.

Voice of Customer (VoC) is a function helping us look at contact centers from the customers' perspective. Though agents can be quite good at following the script, properly greeting customers, and so on, the ultimate answer to the question, "Is this a good contact center?" must ultimately come from the customers. VoC typically administers post-call surveys in which customers answer a few questions and rate the quality of service they received in the contact center. VoC can also be used on alternative channels such as email or SMS—imagine receiving an SMS after contacting a call center, containing a link to the post-call survey which you can complete online from your mobile. We will talk more about VoC when we discuss the customer experience measurement, but for now, it is important to understand that customer feedback is an important way to measure quality in a contact center.

Hive Customer Feedback Management (CFM) interface for building the survey –
HIVE customer feedback management system. ©IST Networks. Used with permission.

Desktop Process Analytics (DPA) is a rather advanced function of the technology running on an agent's PC that "understands" what the agent is doing. DPAs track actions such as launching applications, clicking, moving windows, typing, scrolling, dragging, and so on. With DPAs, quality monitoring is raised a level because, besides call evaluation (the agent *saying* the right thing), we can now evaluate agent actions (the agent *doing* the right thing).

You could say that a DPA is similar to a screen recording, and these can serve similar purposes, such as understanding what the agent is doing on his/her screen, but DPAs are much more effective. With screen recordings, all we have is a video file which we must **physically watch** to understand what happened, while DPAs give us the full history of actions, such as "opening notepad at 11:05.34AM," "minimizing CTI window as 11:05:48AM," "launching Outlook at

11:07:20AM," and "starting to write an email at 11:07:27." DPAs give us a **searchable** record of actions that happen, while a simple screen recording is a video file that one must view in order to understand the content. It is as if you were trying to find the best part of your favorite movie but have to search through it because you don't know exactly where it is. By contrast, a DPA would give you a searchable script, saying, "At 01:35:12, Bruce Willis decides to save the world." The best result is achieved when screen recordings and DPAs are used together. Continuing the analogy above, you could now scroll your video to the exact position at 01:35:12 to see how exactly Bruce Willis saves the world at that particular time. With the combination of screen recordings and DPAs, the result is the same—when we see the event of an agent opening the notepad at a certain time, we can scroll the video file of the screen recording to that exact position to see why the agent needed to launch the notepad application and what he did with it after that.

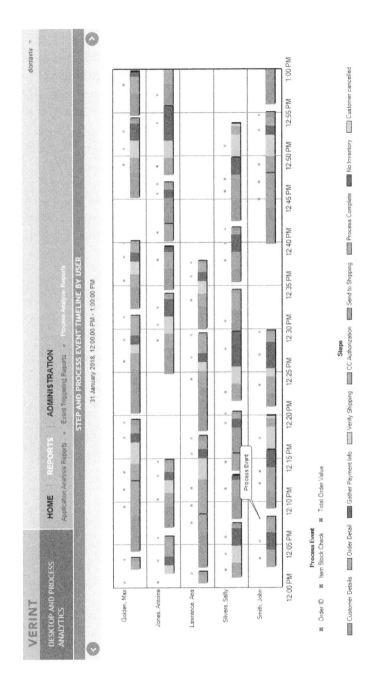

Verint DPA solution screen shows the actions agent performed during a call. Different applications agents use are highlighted with different colors. © 2020 Verint Systems Inc. All Rights Reserved Worldwide. Used with permission.

Agent Scripting is a simple yet effective tool to help agents answering common calls, following an easy-to-use script. You can think of this as a cheat sheet for agents, making it easier to ensure the process of answering typical inquiries is consistent without the need for agents to memorize all possible scenarios. Agents just need to look on the screen, see what they have to say or ask, get the answer from the customer, click "next" to get the instruction that follows, and continue in the same way until the call is complete. Needless to say, embedding the scripting tool into the CTI is the best way to implement it, avoiding the need for agents to switch between application windows.

An intuitive agent scripting tool by Novelvox, embedded into the Cisco Finesse CTI (registered trademark of Cisco Systems) © Novelvox, all rights reserved worldwide. Used with permission

Workforce Management

Workforce management (WFM) is a subset of the WFO, aimed at ensuring the contact center has an adequate number of people to handle incoming contacts effectively (e.g., calls, emails, or chats) according to a predefined criteria (for example, customers must never wait longer than 40 seconds for their calls to be answered). The criteria used by WFMs in different contact centers are usually similar. More importantly, they are measurable. These criteria—often called KPIs—are topic for a separate chapter, but what must be understood here is that answering the calls is simply work to be done, and this work requires people to do it. If work must be done on time—for example, calls should not stay longer than 40 seconds in the queue, meaning that calls should be answered promptly by agents—we must have *enough* people to do it. If we have fewer people than we need, our KPIs will be bad (for example, if calls wait 60 seconds instead of 40 to be answered). The problem is, too many agents are also bad because people are the most expensive part of the call center, and having a shift with 300 agents for a volume of calls that could be handled by 250 will raise questions from the company's management.

This can be illustrated with a supermarket example. I am sure that everyone has experienced waiting in a long line in a supermarket, having noticed that many checkout counters are closed. It would be fantastic for the customers if all counters were manned at all times with employees waiting for us to come by with our shopping carts, but this would be too expensive for the supermarket to pay these people if they have nothing to do most of the time. We can make this example more interesting by using a supermarket open 24/7. In this case, the number of customers at 3 a.m. will not the same as at the 7 p.m. rush hour, so a good strategy is required to predict and allocate the correct number of staff needed at any moment in time, taking into account that during promotion periods, the number of customers will increase, the flow of customers is not the same on weekdays and weekends, and the number of shoppers on Christmas is likely to spike unpredictably. As a cherry on the cake, let's take into account the fact that the average shopping cart size is different at different times. A typical family buying a week's worth of grocery shopping on Friday evening is not the same as a man stopping by at midnight to pick up a bottle of soda and a sandwich for a later dinner (if you had a strategy or formula in your mind by now, that last bit of information probably just destroyed its feasibility).

In the world of contact centers, the challenge of allocating the right number of agents at any given is more difficult. There are multiple, changing factors, dictating how many calls the center gets. Some of these factors are external and even worse, they are unpredictable. Let's look at an example to illustrate why this happens.

Assume a telecom company runs its contact center 24/7. The volume of calls changes during the daytime, typically reaching its peak at 11 a.m. on weekdays. On weekends, the peak shifts to 12 noon. The company has decided to offer a new deal on the last generation of smartphones for its customers, so this will create an additional volume of calls (customers will have questions about the new promotion), and we cannot predict the duration of the calls. The company had some service interruptions in the previous week due to roadwork in some areas, which created a spike in call volume in the evening, as there weren't enough agents to handle the spike, and the queue waiting time increased to over two minutes. In the next week, the company is expected to launch a promotion for new mobile plans, and they expect many customers to call the center to switch to that plan or ask questions about it. The call center manager knows there will be a sporting event the following Wednesday, with a lot of tourists coming in, and they are likely to call the center to activate their new mobile lines (so ten more people are allocated to work on Wednesday), but he does not yet know that the expensive mobile device the company will offer on promotion has a firmware bug, which will cause a massive amount of calls with complaints that Friday evening.

With such a combination of predictable and unpredictable factors, ensuring the contact center always provides adequate service to customers is the job of the contact center manager, which is a nightmare (you can ask them yourself, and I'm pretty sure they'll tell you exactly that). With a workforce management system (WFM), their jobs become a bit less complex because WFMs offer useful functions to predict the load, allocate the resources (agents) to handle the load, and make sure everything goes as planned on a daily basis.

In this chapter, we discuss only a few functions of the WFM: forecasting, scheduling, intraday management, reporting, integration with payroll, and shift trading.

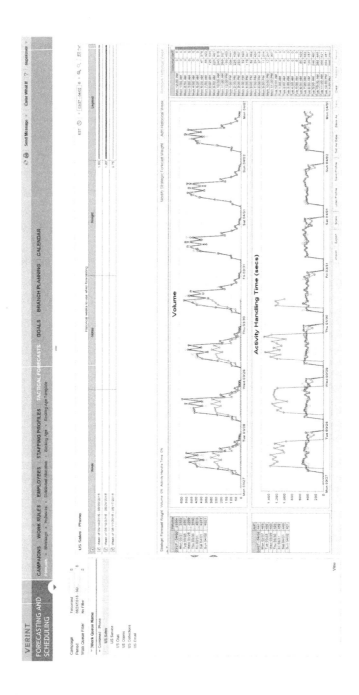

VERINT Forecasting Screen Views. © 2020 Verint Systems Inc. All Rights Reserved Worldwide. Used with permission.

Forecasting is the most important function of the WFM. Based on past data about the call volumes (which any contact center software stores for at least a year), the WFM roughly predicts future call volumes. The contact center manager can adjust predictions with the introduction of extra information regarding external factors, which might affect call volume, such as planned promotions. The ultimate result of the forecasting process is the prediction of how many calls may be received at a given period of time in the future, and which skills will be required to handle the calls—remember that if we expect a large amount of sales-related calls, we have to make sure that agents with relevant sales skills will be available to handle them).

Scheduling is the process of creating a schedule for agents to work, according to forecasted call volumes. Good WFM software understands that all agents are different (some of them may have one skill, some may have several skills), takes into account the maximum amount of time an agent can work per day, the minimum amount of rest needed between shifts, and many other factors. The scheduling process is also affected by the required KPIs the call center has to achieve. In other words, if we must answer calls within ten seconds, more manpower is required than in case it is still acceptable answering them within thirty seconds.

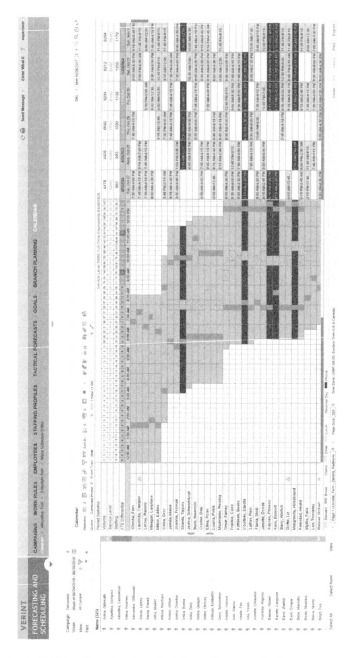

Verint WFM interface showing schedule for the agents in the contact center. ©
2020 Verint Systems Inc. All Rights Reserved Worldwide. Used with permission.

Intraday management is the process of managing the workforce "on-the-fly." No prediction is absolutely accurate, so the schedule of the actual work day may require small modifications—for example, if the contact center manager sees that more agents have been allocated than required, meaning that these agents can be removed from the contact center team and assigned other tasks for the day. It could also be the other way around—someone might get sick, and the contact center may start to experience a lack of resources for a certain skill. In this case, the contact center manager may decide to bring someone in from home or extend a shift, giving an agent a few hours of overtime.

Reporting is an important aspect of the WFM, and it gives the contact center administrator a way to see what is happening in the contact center from a workload vs. available resources perspective (in addition to other reports available from the WFM which is also available from the contact center platform). Good WFM solutions typically offer modules for analytical reporting, so the administrators can analyze the data. This can help them plan operations better in the future.

Payroll system integration is important in environments where agents are working flexible schedules and are paid according to the exact amount of work they do. WFMs can extract data from the payroll system about the actual work agents do, automatically calculating the required compensation for the staff.

Shift trading is the facility some WFM systems give to agents to change shifts. For example, if John usually works the evening shift and Annie the night shift. If both of them serve the same types of calls, these two can agree to swap their shifts. WFMs control the shifts allowed and confirm that the two agents share the same skills so that overall resource availability in the contact center will not suffer from the swap.

Summary and Quiz

In this chapter, we discussed some WFO tools used in call centers. WFOs help us manage contact centers for the benefit of the customers (providing better service), companies (adequate cost vs. performance), and the agents themselves (fair assessment of their performance and work). The WFM industry is still developing, and there are some software companies specifically focused on this sector and many advanced products available. In this chapter, we just scratched the surface. I encourage you to do further reading and research on the topic.

Please answer these self-check questions to test your knowledge of the topics covered in this chapter:

1. Which of the following statements is true?
    a. Workforce-related expenditure is insignificant compared to telecom and technology costs.
    b. Contact centers should hire more agents to avoid long queue-waiting times in order to reduce telecom costs.
    c. Contact centers should invest in technology, which will allow them to utilize the workforce more efficiently.
    d. In today's contact centers, technology cost share is increasing due to technology becoming more complex and expensive.

2. Which of the following is not a typical function of WFM systems?
    a. Forecasting future call volumes.
    b. Playing back recorded calls.
    c. Creating agent schedules.
    d. Allowing agents to change shifts with each other.

3. Which function of recording systems make it possible to choose the right category of calls for evaluation?
    a. Call metadata
    b. Screen recordings
    c. Encryption
    d. Compliance

4. Which two (2) methods prevent unauthorized access to sensitive information?
    a. Masking
    b. Selective retention
    c. Encryption
    d. Evaluation

5. What are the two (2) biggest concerns associated with screen recording?
   a. The large size of the resulting files.
   b. It is typically required for compliance.
   c. It is not usually provided by WFO vendors.
   d. Sensitive information displayed on the agents' screens may be recorded.

6. In an IP contact center, where can the media stream be forked to send to the voice recorder? (choose 3)
   a. Agent's phone
   b. Media gateway
   c. Switch
   d. Patch panel
   e. Microphone

7. Which of the following functions help to reduce the storage size required to keep recordings (choose 3)?
   a. Selective retention policy
   b. Encryption
   c. Compression
   d. Recording only five to ten percent of the calls
   e. Quality evaluation of calls
   f. Live monitoring of calls

8. Which of these functions produce a searchable trace of actions performed by agents on the screen?
   a. Call evaluation
   b. Screen recording
   c. Compliance recording
   d. DPA

9. Call evaluation requires... (choose all that apply)
   a. A screen recording be associated with the call.
   b. An evaluation form must first be created.
   c. Recorded files to be present in the system.
   d. Agent eLearning modules.
   e. VoC modules.

10. In a small contact center working 9 a.m.-6 p.m. with only one fixed agent shift of ten agents, adding which of the below systems provide no value?
    a. Compliance recording
    b. Quality management
    c. Workforce management
    d. Screen recording

Day 4: Outbound and Multichannel

In the previous chapters, we discussed how contact centers process large volumes of incoming calls, but phone calls can be initiated in two ways, and so can customer communication. We mentioned multiple use cases in which customers call the call center to place an order, solve a problem, request a service, or get some information.

In this chapter, we extend that knowledge and discuss outbound scenarios in which contact center place calls to customers instead of waiting for the customers to call. We also learn about multichannel contact centers in which customers can reach contact centers not only via voice calls but also via something known as "digital" channels: chat, email, and social media.

Outbound

We have already discussed multiple situations in which customers need to contact a company to request a service, file a complaint, order something, or get information. These are called *inbound* scenarios because the customers are the ones calling, and the contact centers receive the calls. Outbound scenarios—in which call centers call the customers—also a number of purposes. Here are a few examples of outbound use cases:

- selling goods or services to new or existing customers (sales campaigns);
- collecting debts and payments (collection campaigns);
- notifying groups of people about certain situations, such as emergencies or outages;
- *proactively* updating the customers about the status of their orders or requests ("proactively" because call centers call customers instead of waiting for customers to call to ask for the information);
- collecting information (social survey campaigns, or confirming customers with scheduled appointments are going to attend);
- improving or measuring the overall quality of customer service (for example, calling customers back if they have left a long,

inbound queue, or calling customers back who have requested a call-back via IVR, web site, or mobile application); and

- collecting information about customer satisfaction with the company (or call center), services, or products.

There are many ways to organize outbound call centers (from a process, people, and technology perspectives). As it often happens in the call center industry, different vendors call the same thing by different terms (or the same term might mean something different to different vendors). To make it simple, I will use Cisco's terminology, where it comes to outbound systems. The other two industry leaders—Genesys and Avaya—are a bit different, but the ideas are always the same: to organize outbound calling activities optimally.

Let's focus on the outbound from the three aspects mentioned above: **the people, process, and technology**.

People (or agents) are, as we already know, the most expensive resources in a call center, and the most important aspect that call center management has to address is if the same or different agents will be used for both inbound and outbound calls. If the same agents are used for both inbound and outbound operations, this is called **blending.** The choice to use agent blending or not depends on factors dictated by the business. For example, in a bank, the incoming call center is used to help customers, answer inquiries, and so on, while the outbound calls are frequently used by the collections division. The collections division requires completely different call-handling skills from the agents and most—if not all—banks I have worked with prefer to separate the collections division logically with a dedicated staff, but if outbound is used to call customers—who were waiting in the queue but couldn't reach an agent—back, then it makes perfect sense to use the same team as the one used in the inbound call center. One more aspect to consider is the volume of outbound calls that need to be considered by the workforce management system. For example, a company may choose to start a massive sales campaign, which requires the calling of a few million potential customers, and for this purpose, they may temporarily hire some part-time agents. I have even seen cases where such campaigns are large enough to warrant call center management contracting such campaigns to outsource contact centers. We will discuss the concept of outsourcing the outbound again when we reach technology aspects later in this chapter.

One case we should mention here is when people are not used at all in the outbound call center. In this scenario, the IVR is used to play notifications to customers automatically. This may not always be the best way to announce information to customers, but it is acceptable in some cases, and technology offers this possibility in most current contact center platforms.

The **process** is the second most important aspect to plan in outbound. Outbound is about calling X number of customers. Here we define how to decide *who* we are calling (which must include phone numbers), how calls are *distributed* to agents, how *personalized* the calls are (will we deliver customer-specific information or are we calling tens of thousands of customers and try to sell them the exact same product?), if agents will need to review customer-specific information before they talk to customers, if agents need to collect or record information during the call, and if there is the potential possibility for the call to be dialed again.

A well-defined process is usually easy to convert into proper (and more importantly, adequately priced) technology, so let's use an example to illustrate this.

A bank decides to offer a free gold VISA credit card to all of its customers, maintaining an average balance of $4,500 over the past three months. All customer details and histories are available in the banking system, so a list of eligible customers can be generated, which will be done every week. Then, the list is filtered to exclude customers already possessing gold VISA credit cards or any other VISA cards with a higher privilege level, as it would be silly to offer a product to customers who already have it (though I personally had many companies calling me to offer a service I already had, meaning that they did not design the process well; it also means they are losing massive amounts of money due to the poor design). The campaign runs every day between 1:00 p.m. and 3:00 p.m. If customers are interested in getting the card, agents send an update to the banking system flagging the customers for the credit card department, which is then responsible for issuing the card and delivering it to the customer. The agent's responsibility ends after he flags the customer as being interested in the banking system or not. Before closing the call, the agent asks if the customer is interested in receiving any other services or information. If the customer is interested, the agent transfers the call to the inbound call center with a high priority, so the customer does not wait in the queue for more than five to ten seconds.

To complete the story, it makes perfect sense to record the *sizing* information. The expected number of calls every day is between 3,000 and 4,000. If a customer picks up, the average call duration is between three and four minutes; we expect that not more than 30% of the customers will answer the call. The sizing information is important when choosing the adequate technology. If you call ten to 20 friends for a party on Friday night, you can do that manually. If a call center needs to notify 1,500 customers daily about their credit card expiration, automating it will help and is worth paying for. When there are campaigns requiring tens of thousands of numbers to be dialed, the proper outbound technology is a must.

This brings us to the discussion about technology and elements of the outbound system:

Campaign—usually defines a kind of a logical boundary for a reason to place outbound calls. For example, a campaign to notify passengers that their flight will be delayed. Campaigns can have multiple properties associated with them from a technical perspective (such as a calling list, dialing mode, and necessary skills the agent must have) and a business perspective, which—as in the example above— can mean a campaign for notifying all passengers of all flights delayed at a given time. Alternatively, the contact center may have different campaigns for domestic and international passengers if the business is required to separate them for any reason.

Dialing list (calling list)—a list of phone numbers (usually associated with names) that call centers need to dial.

Dialing mode—the way numbers are dialed. For example, displaying customer information on the agent's screen before connecting a call or the number of lines dialed per agent line (for example, if you want to keep ten agents busy, you may choose to dial 15 numbers from the list because you expect that some people will not answer the phone). We will talk more about dialing modes later in this chapter.

Campaign manager—a module that "understands" campaign properties and manages the entire process and all modules, including the dialer. Campaign manager software has become more advanced and allows combining classic phone outbound calls with reaching the customer through other channels. For example, after three attempts to reach a customer by phone, the campaign manager may be configured to send him/her an SMS or email message.

Dialer – a contact center module responsible for dialing numbers so agents do not have to dial them manually from their keypads or phones. Dialers are managed by the campaign manager, which tells the dialer which number (or set of numbers) to dial.

Do not call (DNC) list—a list of phone numbers which should not be called under any circumstance. Usually, the DNC is applied to all campaigns.

There are different scenarios behind how the process of dialing is organized in outbound call centers. The first scenario is where the calling list is somehow generated or uploaded to the system, usually as a text file in CSV format, which has been generated using *Excel* or similar software. Campaigns are configured by administrators, and campaign managers start the dialers, which dial the numbers in the list and assign calls to agents. After calls are connected to agents, they are removed from the campaign, and the rest of the process is completed by the agents. Agents may still choose to return calls back to the campaign (for example, if the customer is interested in the product but may ask for someone to call again later).

The second scenario is sometimes found in call centers in which the role of the campaign manager is played by the CRM (customer relations management) system. If you have never dealt with a CRM, you can think about it as a customer database, storing customer details—such as addresses, phone numbers and *relationships* between companies and customers—for example, products customers have purchased, complaints they have raised, and customer statuses (for example, VIP customers). In the second scenario, the CRM decides to give to a particular employee the task of calling a particular customer (or the employee chooses this from the CRM record as a part of his job). Then, the employee uses the click-to-call functionality on his PC (integrated with the phone system) to place the call to the customer.

The difference between scenarios one and two is how the calling process is implemented. In the first scenario, the dialer dials based on the list. The list may still come from the CRM system, but after it, the

CRM has no control, which is carried out by the campaign manager, and the dialing is done automatically by the dialer, and calls are sent randomly to agents based on skills and agent availability). In the second scenario, the *CRM controls each and every record*, and the agents dial the numbers using click-to-call in a one-by-one way. The first scenario can have a CRM in place (and usually does), but the CRM is displayed to the agent when the call is connected (or just before it), and the entire process and selection of which number to dial is done by the campaign manager.

There are several ways the dialer works, called **dialing modes**, dictated by a combination of performance vs. the level of attention the call center wants to give to each of the customers. Remember that agents are the most expensive components of the call center, and if we have a team of 20 agents in the outbound division, for example, our job is to utilize them wisely. Not all people answer the phone, so if a dialer makes only 20 calls, there will be some calls unanswered, and some agents will remain idle. You may need to make 30, sometimes even 40, calls to have 20 calls successfully connected (and thus utilize our workforce of 20 agents efficiently). The trick is to know the exact number because if you have 20 agents but 21 customer answers, the 21st customer will be put in the queue. This is something you should avoid as there is nothing that frustrates a customer more than receiving a call from a call center and finding that he's been placed in the queue; I find it disrespectful and simply hang-up. Dialing modes define how we balance between performance (keeping the maximum number of agents busy) and avoiding new calls landing in the queue.

Predictive mode is when the dialer dynamically predicts the ratio of phone calls answered by customers, and it places the outgoing calls accordingly. For example, at any given moment, the dialer may sense that the **answer rate** is 40%, meaning that four calls have been answered from ten. This, in turn, means that we need to make approximately 50 call attempts to keep our agents busy (out of 50 calls only 40% connect, which is exactly 20). The dialer may be configured to make slightly fewer calls to err on the safe side (it is better to deliver 19 answered calls rather than 21). Predictive mode is good for large outbound call centers where a large number of records—hundreds or thousands per hour—need to be dialed. Because it adapts to the changing answer rate, it requires fewer adjustments. Because of the statistical calculation method used by the predictive dialer, it is usually not the best choice for small outbound operations with fewer than 30-40 agents.

Progressive mode is similar to predictive mode but the actual prediction of how many lines to dial is done by the system administrator. Progressive mode is useful for medium-sized outbound operations with a minimum of ten to 15 agents. Unlike predictive mode, it does not use statistical mechanisms, it is not prone to errors, but it still allows a relatively large number of calls to be placed quickly—much quicker than with the preview mode.

Preview mode is when a pop-up containing information about the call is displayed in front of an agent, and the agent accepts the record. Only then is the phone number dialed by the dialer. Preview mode is the slowest of the dialing modes in terms of the number of calls that can be completed over a period of time because previous modes *allocate the agent first.*

Predictive and progressive modes do *not* allocate the agent first—they place the calls *knowing* there are X free agents available (or becoming available every minute), so they keep dialing, and you cannot dictate which agent gets which call in the campaign. With preview mode, the dialer dials the number when an agent has already been allocated, assigned, and accepted the call.

Despite the fact that preview mode is much slower, it is useful in performing a more *personalized* service for customers. This is because the agent can review customer records and prepare for calls before they are placed. Also, preview mode always grants that a call will be attended by an agent, while predictive and progressive dialers may run into a situation where the number of answered calls is greater than the number of available agents.

Scheduled callback mode is when a customer answers a call but asks the agent to call him back later. The agent then returns the call to the campaign, and the dialer places the call at a time requested by the customer (or as close to that time as possible), optionally delivering the call to the same agent (known as **personal scheduled callback mode**).

As you can see, organizing the outbound division of a contact center requires understanding and careful planning. Moreover, it can potentially require an investment in people (hiring) and definitely in technology (buying licenses for outbound systems, which are usually sold separately by contact center vendors). In many cases, it makes better sense to *outsource* the entire outbound operation to a third-party provider. Examples are some seasonal, irregular campaigns that need to take place at a certain time of year, and are not active at any other time. For example, airlines promoting a sales campaign close to vacation season to attract the maximum number of customers.

It typically does not make sense to hire personnel, install (or upgrade), and configure equipment and software for something required only for a few weeks a year, so outsourcing is the best choice for this.

Multichannel

Phones have been around for a good number of years, but the Internet has introduced new ways of communication, such as email and chat messengers. Mobile networks also come with SMS, a simple yet effective way of communicating. All of this has mixed together over the past ten years—messengers support voice calls, web sites have integrated chat options, and Web2.0 with its social networking and complicated, feature-rich messaging tools. For Generation-Y (those born after 1980), this multitude of communication tools has contributed to the quick demise of strict phone communication as the sole means of talking to one another.

Today's companies quickly realized that serving customers over the phone alone is not good enough, and they started to adapt their contact center technology to allow for communicating with customers via the channels customers prefer to use. This technology has created the *multichannel* contact center because it offers customers different ways with which to make contact. We discuss a few channels and their specifics here.

Web chat – when a customer initiates a chat session with a contact center agent from the company's website or smartphone application. The interactive form may require the customer to input information such as name, customer ID, contact information, and the type of request he/she has. This information can be useful in identifying which agent should answer the request using algorithms for skill-based routing. Web chat requires customers to keep their browsers open for the duration of the chat; therefore, the chat session is usually limited to ten to 15 minutes. On the contact center side, one agent is able to chat with several customers at the same time, and depending on his or her abilities, they might be able to handle four or five conversations at a time. Web chat is a useful channel for businesses providing services via their website (e.g., airlines, online stores, booking engines) where customers might require assistance to help finish transactions (such as buying tickets). Web chat has a few useful functions to facilitate customer support, such as:

- **Co-browse**—where the agent might be able to see the customer's desktop
- **Callback**—where the customer may request the agent to call him/her back to facilitate communication
- **Chat transcript emailing**—in which the customer receives a complete transcript of the chat conversation he/she just had, which may be useful in some cases
- **Video or call escalation**—in which the chat is converted to a phone or video call in the same session
- **Canned responses**—typical phrases agents can quickly insert into a chat instead of typing them manually every time. These include standard greetings, responses to common questions, typical phrases, and so on. An example of a typical canned response might be, "Thank you for chatting 123Flowers. My name is Charly. How may I help you?"
- **Knowledgebase integration**—a function in which common problems, inquiries, and their responses are already documented for the agents to re-use to send to customers. For example, a telecom company may have standard instructions on how to activate Internet connections for different models of mobile phones, so the agent can quickly push the right instructions to the customer when needed.

Mobile chat—basically the same communication channel as webchat, but instead of a company's web site, customers use the mobile app. This method often makes it easier for customers because they have already been authenticated in the mobile application, and information—such as their customer ID, name, and contact details—can automatically be pre-populated in the chat form. Besides that, there is no longer any need to have a browser session open because the browser is not used anymore.

Email—a powerful way to communicate with contact centers because customers may take their time to describe situations and attach data—such as images or documents—to send to the company. Emails are also routed using skills-based routing, and the information used to route emails may include the email address of the customer and the email address to where the communication was received. For example, emails received at the support@company.com account will be routed to agents with support skills. Keywords in the subject line or the email body itself may also be of use. For example, a company may ask all inquiries related to the iPhone-11 sent to the support@company.com address with "iphone11support" in the subject line or a "#iphone11support" hashtag in the body of the email. The system spots the keyword and automatically routes the email to the agent group with the skills to answer the inquiry. Unlike web chat, email communication can span many days or even weeks if the cases are complex. Email systems in contact centers usually track the entire email chain as something called "case," from the beginning until the email chain is finalized and the customer gets the information or service needed. This means the case may be handled by multiple agents. Email contact centers usually have the technology for the agent to view the entire case history before replying to a customer. Email systems usually have an agent toolset similar to that of chat systems, including canned responses and knowledgebase integration. Email systems usually have an "auto-response" acknowledgment at the system level, something to notify customers that an email has been received and will be answered by an agent as soon as possible (similar to the way an auto-attendant does this for a phone call).

Social Media Contacts—communication with customers via social media posts and their comments. This could easily be the topic for a separate book, and there are, indeed, many books written about social media customer care and social media marketing (SMM). You might ask why we are discussing it here at all (and moreover, why contact centers should be responsible for social media customer care when social media is often handled by another department in the company). The biggest reason I see for social media customer care being a part of a contact center is that contact centers (strictly speaking, call centers) have been there for a good few decades, and over this period of time, the call center industry has developed a set of practices and tools to effectively manage customer contact handling. Think of social media as a communication channel a little different than a voice call, chat, or email, but in the end, it is still a communication channel used by customers to communicate issues and feedback and receive service. Indeed, if one is not happy about one's in-flight service, he or she can contact the airline via phone, Facebook, web chat, or visit their branch in person—the issue remains the same and should be received, documented, and handled the same way. Still, there are multiple reasons as to why social media customer care should be handled by a separate department or, at least, by a dedicated group of people in the contact center or customer care division.

Reason number one: requests received via social media are often very different than common requests received through other channels, such as phone or chat. Phone calls can be convenient for ordering pizza delivery, while web chat can be useful to get information about the best home insurance plan. Email can be useful for describing a complex problem in the corporate IT system and send it to the help desk contact center when the use of social media in this and the other scenarios might be questionable. We could argue that the channels the contact center already handles are different in nature, and the nature of requests is also different, but they still function well under one umbrella.

The second reason is more serious—most social media communication happens in the public domain. When a customer leaves a comment on a company's page on Instagram or Facebook, and the company replies, both comment and reply are potentially seen by thousands of people. This kind of interaction requires more attention because of the impact it can have. Comments from a highly influential person (a person with thousands of subscribers) should be handled quickly. Negative comments should be handled more quickly than positive ones. Question comments (i.e., comments to the answer of a question) are potential revenue leads (for example, a company publishing a picture of a new model of car and someone asks about its safety features) and should be answered both quickly and consistently.

This points out an important dilemma. Social media posts need the same intelligent engine to prioritize them (by influence, by sentiment—negative, positive, neutral, or a question—and by the type of product or service), and this kind of engine is available in contact center platforms as another kind of skills-based routing. Due to the nature of social networks, every company's answer may be seen by thousands and easily become an official message. Social media customer care must be controlled by the marketing team to ensure the message to its customers is consistent. We are not saying that marketing should not influence the contact center's other channels, such as phone or email (it definitely should), but when it comes to social media customer care, this influence and control becomes more critical because giving the wrong answer to a customer in front of thousands of other customers can easily cost a company with respect to its reputation.

One more reason why social media is often handled by a department separate from the contact center is that companies tend to implement it (or think they implement it) along with social media marketing (SMM). SMM is more than just answering customer comments on a social network—it's about maintaining the company presence on social networks, pushing new, quality content onto the company's pages, reaching out to customers, and so on. Again, the content pushed onto social media requires comments and customer inquiries to be answered on the same channel, so the reason for separating this function from the contact center brings us immediately back to the reason for having these two together.

The final answer, therefore, depends on the company's business model, the company's marketing—and specifically, social media marketing—strategies, the capabilities of the products, and the type of social networks on which the company chooses to maintain its presence, the company's internal structure, and many others.

A separate discussion should be held around messenger apps, which can be a part of the social platforms or can be platforms of their own.

Messenger applications—such as WhatsApp, Facebook Messenger, Telegram, Viber, and to a certain extent, Instagram direct messaging and Twitter direct messaging—have become communication channels so popular, they are impossible to ignore. There are few specifics of these channels that we must keep in mind: they are ever-changing, different from the commonly-known, traditional channels (voice, chat, email), some of them are associated with their "parent" social channels (Facebook messenger with

Facebook, Instagram direct messaging with Instagram), and each of them has its own unique set of features and capabilities that may create additional issues when handling them. For example, a Facebook messenger conversation can be converted to a voice or video call. The same is applicable to WhatsApp but not Instagram direct messaging. As a "bonus," to make the situation even more complex, some vendors don't allow integration of their messengers with external platforms (including contact centers).

Because of this, a lot of companies with strong, well-established contact centers find themselves completely lost and unprepared when it comes to adding messenger channels into their customer care environments. Many choose to wait rather than invest in technology that may completely change before the integration work has been done, and the channel is officially launched as a part of the contact center.

It is not only the technical integration but also the control and measurement that has to be largely reinvented when dealing with messenger channels. Voice conversations have simple metrics—such as "average talk time"—between agents and customers, and there may be something similar for web chats. The question is, how we can measure it for WhatsApp when conversations may last for weeks or even months because there is no concept of start and finish for a conversation in a WhatsApp channel. And how do we allocate the resources? If one agent can answer one phone call at a time—or three or four web chat conversations—how many simultaneous direct messaging threads on Instagram can he or she control? And how do we measure agent load and productivity and perform forecasting and scheduling? We will discuss these challenges on day six, but we can say that measurement KPIs and mechanisms for social messaging channels are still in their early stages of development, which leaves a good space for creativity and innovation from both contact center experts and contact center technology companies.

Omnichannel – blend it together to rule them all

Nowadays, most companies have implemented some kind of multichannel contact center with sets of communication channels according to their strategy and customer preferences. Multiple studies are available about channel preferences of different customer segments and for a variety of inquiries. Though results can vary across regions and businesses, there is one function customers always expect, but companies rarely manage to properly implement in their contact centers: the ability to switch between channels without the need to "start again."

Imagine a perfect world in which you must book an airplane ticket using an airline's website. At some point, you will need help, as you are concerned about short transit times at the airport and would like to double-check with the airline that they'll be able to move your baggage between flights in the less than an hour you will have. You click a button to start a chat conversation and the agent that joins the chat with you after a few seconds already knows that you are booking your flight from Singapore to Milan via Istanbul, your dates, your flight numbers, and he is ready to answer your question immediately (without having to ask a million irrelevant questions that, as we all know, agents love to ask and we love to answer including—but not limited to—the maiden name of our dog's mother). Imagine that your question is quickly answered, and you are going to finalize your booking. You are on the payment screen, but the payment does not go through as the application does not allow you to choose a certain field for your credit card. You request a callback via the web site and receive a call from an agent in a minute or so. The agent already knows your name and the flight you are booking and helps you to finalize it right away. This is shown on the screen as you talk. All that is left if for you to press "confirm" to complete your payment.

This example illustrates a customer care system in which all channels (web site, call center, web chat, and perhaps others) are integrated with each other in a way that allows you to switch between channels without the need to start again. We started on the web site, had a chat, continued on the web site, got help via phone, and returned to the web site to finalize our transaction. There can be multiple *customer journeys* and different combinations of channels. *Omnichannel is a concept in which the contact center's system has its communication channels integrated in such a way that switching between them is possible without the customer needing to repeat the information or start the conversation or transaction from the beginning.*

Omnichannel is not a module, product, or system. Rather, it is a strategy and a vision that requires the following elements to be considered and implemented:

- **Customer data**—we must know the customer's information such as name and contact details such as phone, email, customer ID, and as in the example above, perhaps a frequent flier number. This information helps us identify the customer and build a strategy around him/her.
- **Context information**—a record of what the customer is trying to do, for example, booking a ticket to Milan. Unlike customer data, which is mostly static, the context information is always dynamic and is the second key to serving a customer "right here, right now."
- **System integration**—the ability of contact center systems and channels to "talk" to each other and exchange information about customers in near real-time.
- **Customer journey**—a concept of "typical" scenarios concerning multiple channel and system usage that we track and catering to. For example, visiting a web site, searching for a flight, choosing a schedule, choosing options, adding information, proceed to payment, and receiving a ticket via email is an example of a customer journey. We may have more complex journey scenarios, such as the one above, in which the customer had to contact the airline twice before the ticket was booked.

Omnichannel strategies require understanding (and monitoring, if possible) of customer journeys and the enabling of the necessary channels, context systems, and building integration between them to make the customer journey as smooth as possible. Implementing omnichannels without a proper understanding of customer behavior is likely to be less efficient and not as successful. Omnichannels are easier to build when system components come from a single vendor— or ecosystem of vendors—in which the compatibility has already been tested and confirmed, and even in that case, it is often a long and complex exercise.

The resulting customer loyalty with those who enjoy dealing with consistent, efficient, and easy to use systems is worth the investment. We will discuss ways to measure customer satisfaction and loyalty on day 6, but before that, we come back to old good voice channels and discuss why, despite all the innovation in the industry, voice channels are still the most important to use. We will also discuss how we can make it better and more convenient, given today's *speech and language technology.*

Summary and Quiz

In this chapter, you learned about the difference between traditional call centers and contact centers with multiple channels, integration between these channels, and their characteristics and features that help contact centers to serve their customers better.

Please answer these self-check questions to test your knowledge of the topics covered in this chapter.

1. Which one of the following scenarios does not require an outbound dialer application?
   a. Credit card debt collection departments.
   b. Agents collecting information via survey.
   c. Call centers offering mobile plan upgrades to all existing customers of a mobile operator company.
   d. Agents needing to place outbound calls to customers to get additional information.

2. Using the same agents for inbound and outbound activity (choose 4):
   a. is called "call blending."
   b. allows a more efficient utilization of the workforce.
   c. is technically difficult to configure with most call center vendors.
   d. requires agents to have relevant and related skills and knowledge.
   e. does not make sense if large outbound campaigns are planned.

3. A business scenario defining outbound dialing rules, how calls will be placed and when, etc., is usually called:
   a. an outbound dialing profile
   b. an outbound business template
   c. a campaign
   d. CSV files

e. a do-not-call (DNC) list

4. Which system may, in some cases, play the role of the campaign manager?
   a. CTI
   b. CRM
   c. Predictive dialer
   d. Proactive dialer
   e. ACD

5. Agents are supporting customers via chat. You discover that a large portion of requests served by the agents requires typical answers. You don't want the agents to spend time typing these answers out again and again. Which chat feature(s) can help? (choose 2):
   a. Configure and instruct agents to use canned responses, also known as response templates.
   b. Instruct agents to use call escalation to quickly explain matters to customers in a call instead of typing responses into the chat window.
   c. Use the chat transcript emailing feature, ask the customer to terminate the chat, and then include the typical FAQ in the email transcript.
   d. Instruct agents to use the knowledgebase to make sure common scenarios are addressed there.
   e. Use co-browser features to guide customers to the web-portal where their typical questions will be answered.

6. Your colleague claims that in the contact center he works for, customers can quickly jump from website sessions to chats and then to a voice call or request a callback while agents solve their issues. This type of contact center is called:
   a. Multichannel
   b. Omnichannel
   c. Cross-channel
   d. Multi-modal
   e. Digital

7. What are the three (3) ways to classify and prioritize social media contacts?
   a. By influence.
   b. By size.
   c. By date.
   d. By sentiment.
   e. By product or service.

8. What is the challenge(s) associated with integrating contact centers with social messenger applications? (choose all that apply):
   a. Each of the social messenger apps has its own characteristics and features.
   b. Control and measurement KPIs are different from traditional contact centers.
   c. Technology changes quickly and unpredictably.
   d. Each of them has its own integration specifics or does not allow integration at all.
   e. Some contact center systems have difficulty handling social messaging channels.

9. Which elements are required to build an omnichannel experience in the contact center? (choose all that apply):
   a. System integration between contact center modules so they can communicate with each other.
   b. Dynamic context must be maintained.
   c. Customer profile data should be available.
   d. Customer journeys must be understood and taken into account.
   e. None of the above.

10. What are the advantages of having an omnichannel contact center? (choose all that apply):
    a. Fewer technology costs as they always come from a single vendor.
    b. Contact centers can resolve issues quicker.
    c. Customer experience is more consistent.
    d. Monitoring contact center KPIs is easier as all channels can be monitored through a single dashboard.

e. Workforce management systems can better predict omnichannel loads.

Day 5: Speech and Language Technology in the Contact Center

Speech and language technology is one of the "coolest" topics and not only in the contact center world but in our daily lives. Apple's SIRI, cars with voice-controlled functions, and lifts trapping you inside until clearly announce your desired floor with the right accent, and even voice-enabled search engines—these are all a reality today. Sometimes, I think that my dishwasher will soon understand me better than my own children (and then definitely behave better).

The contact center industry is, to a great extent, affected by speech and language technology, and it also drives its development simply because a contact center is all about customer communication, which depends upon speech and language.

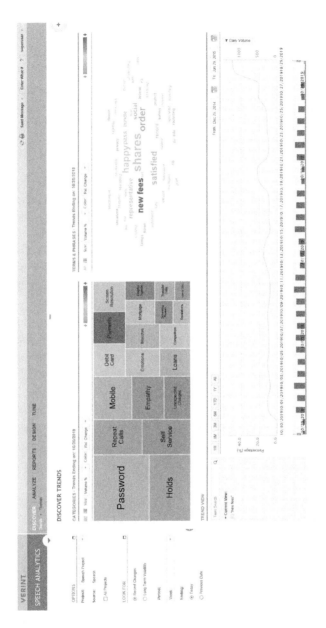

Verint speech analytics system automatically analyzes the conversations in the call center highlighting the most popular topics discussed and keywords mentioned. © 2020 Verint Systems Inc. All Rights Reserved Worldwide. Used with permission

There are multiple vendors and products readily available in the market, all of which are built around a few core capabilities:

- Understanding of how text corresponds to speech. All of the rules for proper reading, correct pronunciation, accents, and dialects provide the scientific foundation for technologies to convert text to speech (TTS) (which is how "computer can speak") and speech to text, also known as automatic speech recognition, or ASR (where we speak, and a computer interprets this into written words).

- Understanding of the language model, which, simply speaking, refers to understanding the meaning of a phrase. This goes far beyond simple keywords, because when we ask, "Do I need an umbrella tomorrow?" the real meaning is probably, "Will it rain tomorrow?" In the ideal case, the system should provide us with the weather forecast. Sentiments—such as sarcasm—can be tricky for the system to recognize. Writing something like, "After two weeks, my AC works so perfectly, I have to bring the car to your lovely workshop again," provides an example where the keywords "perfectly" and "lovely" have the complete opposite meaning.

- Understanding how the voice of one person is unique to that person and different from any of the other five billion people living on our planet is the foundation for voice biometrics, which can be used to check a customer's identity over the phone in the same way an eye scan or fingerprints are used to access systems and verify one's identity.

There is one more key concept we must understand before dealing with any kind of speech or language products: terms such as "success," "failure," "different," and "identical," which are always fuzzy. This means that no system can understand 100% of human language because even human beings often fail to recognize someone by voice or detect sarcasm or irony (frankly, some people are incapable of doing that at all). Similarly, when a system analyzes the biometric profile of a person's voice (often referred to as **voiceprint**), it can return with 99.9% positive assurance that the given voiceprint corresponds to a given person (usually we say that a 99.9% probability is considered a **match** between the two voiceprints). Computer-generated speech can be close to human speech but not completely so—you can still hear a bit of a robotic "accent" in the speech, especially if you know the speech is computer-generated, but this does not make the speech sample less understandable (I prefer TTS-generated speech to the speech of many humans).

In all cases, it is us—people—who decide if a particular technology or product is adequate to complete the function we need it to.

Let's look at some of the most influential technologies and their use, one by one.

Text to Speech (TTS)

In simple terms, text to speech is a system that converts written text into human speech. TTS can be further classified into **dynamic (on-line)** and **offline TTS**. In contact centers, **dynamic TTS** systems are usually integrated with IVR and serve the purpose of announcing, through IVR, information that could not be predicted or constructed in advance using standard sound files. This means that if you call an airline's call center to get your flight status and hear the message, "Dear Mr. Frank Thompson, your flight to London Heathrow today will be on time at 19:05 from Terminal 1," then everything you hear, except for "Frank Thompson," is *not* TTS because you can easily pre-record the sound files for numbers, the word, "Terminal," destinations (such as "London Heathrow"), statuses such as "on time" or "delayed," salutations (such as "mister," "missus," etc.). Because an airline's call center is usually limited to a few dozen typical announcements containing a limited set of information—such as destinations, times, and so on—recording sound files (they are called "voice prompts") for these announcements can be done in a studio, and the IVR system simply strings them together to make a sentence.

Pronouncing "Frank Thompson" is much trickier, as you cannot pre-record all of the customers' names, so if we need that level of personalization, we must use an online TTS to generate sound for us on the fly—in real-time. We can then announce the passengers' names as a part of the longer sentence. To make it sound seamless, the voice of the machine must be indistinguishable from the voice in the pre-recorded prompts, or the experience will sound like patchwork.

This brings us to the use case of an **offline TTS**. Let's imagine that our airline has just added another flight to Kuala Lumpur and needs to upload the corresponding prompt file to the IVR system. This means we need to contact the studio to call the same person who recorded the rest of our prompts (the **voice artist**) to visit the studio to record the new prompt for Kuala Lumpur. This is usually expensive (calling someone in, even for a small task, adds a lot of administration overhead), needs additional planning (as the person may be on leave or can be sick at home with his or her voice in sub-optimal shape), or sometimes impossible (the person may have left for good or even died). An offline TTS can be an instant solution to all of the above problems, as one may generate the prompts on the same day, using a simple software typically running on a PC or online. It is usually simple: you write the text, the system generates the sound, you listen to the sound, and if all is good, save it as a file to be uploaded to the IVR. You may fine-tune the output of the TTS by adding a question or exclamation mark, commas, or special symbols to make the speech faster or slower, increase the pitch, and adjust it in many other ways to sound more natural for the context in which it will be used.

TTS systems are so mature nowadays that a good quality, computer-generated voice is difficult to distinguish from a real human voice, but most importantly, the quality of a TTS-generated voice (in many popular languages) is easily understandable because when we call a call center, all we need is to hear messages that are easily understood. Many of us do not call to measure the quality of the TTS, so a tiny bit of computer "flavor" in the IVR's speech is totally acceptable.

One more important aspect of the TTS system is the voice. TTSs consist of two main blocks: the language model (which is a combination of rules for pronouncing speech and is unique to every language, including accents and dialect), and the voice itself (which, like any human voice, is unique to a person). Normally, TTS vendors offer a choice of several male and female voices for each language, and it is common to name them with human names, such as "Lisa," or "Michael." While language models for each language are the result of years of research and development work, building the actual voices is easier. All that is needed is to have the "original" voice from a real human, and then "train" the generation algorithm to pronounce the text using that particular voice. Not all languages have language models as mature as the English language, for example, but the industry is constantly developing, and I can easily predict that the situation will be completely different by 2025. Even now, you can google TTS engines on the Internet, generate a voice, and see how it works. There are plenty available in common languages that you can test for free to experiment with how the technology works.

Automatic Speech Recognition and Natural Language Understanding

Automatic Speech Recognition (ASR) is computer software that can recognize human speech and convert it into a written text. There are many "levels" of maturity (or complexity) that an ASR system can have, and it really depends on the understanding of the word "recognition." There are systems able to match human speech to a number of pre-defined (and expected) words, systems able to understand almost any word in a language dictionary, and systems able to go beyond that to automatically understand the meaning of a sentence. Each of them has its own use cases, and we will cover two cases here: limited dictionary ASRs and complete **natural language understanding** (NLU), powered by an **artificial intelligence** (AI) engine.

Let us go back to the airline contact center example used in the previous section, where customers may call the airline to get the flight status or book a ticket. As a part of the IVR flow, the customer may need to input the flight destination, at some point. For any airline flying to more than three or four destinations (which is most of them), reading a list beginning: "press <01> for London, press <02> for Paris, press <03> for Chicago," and so on is not an acceptable solution when there are dozens—if not hundreds—of destinations in the network. In this case, installing an ASR system trained to recognize customer voices (or asking the software to "map" what a customer has just said to an entry in the list of pre-defined entries) can be a great solution. All that is needed is an ASR system that can be "trained" (kind of a "fine tuning" of the system) to recognize destination names such as London, New York, and Chicago, and integrate this into system with the IVR adding the message for users to "Please say the city of your flight's destination."

If a customer says something random—for example, "A city of fashion"—the system simply does not recognize it and asks to repeat the attempt.

The experience is different with NLU because the IVR's menu structure may not be needed anymore. All that is needed is for the IVR to say, "Dear customer, please tell us what do you want to do." The customer responds with something like, "I would like to book a ticket to the city of fashion" (I would still suggest defining cities by name unless you want to fly to a random destination). The IVR then confirms your choice: "Do you want to book a ticket to Milan?" (or likely, first asks you to specify the city by name), and after you respond, "Yes," it activates the step by step process of booking a ticket via the IVR asking you for dates, names, and other necessary details needed to issue your ticket.

The ASR- and NLU-powered IVR no longer needs menus—the whole structure is as simple as, "Please tell me what do you want us to do," and the computer's ASR/NLU engines interpret the speech, analyze it, and map it to a list of pre-defined intents. By intent, we mean functions the system can perform or understand that the requested function is not available. In the latter case, the IVR has to "apologize" to the customer and transfer the call to an available agent. Thus, if an IVR supports booking tickets, getting a flight's status, cancelling/changing tickets, and you suddenly ask it to issue a new frequent flier card, a properly designed system should understand that this is not one of the functions the system normally performs and transfer the call to an available agent for assistance.

The perfect system described above consists of two modules (or layers). First, human speech is converted to text and analyzed by an AI engine to understand the actual meaning. This is not a keyword analysis but a complex process which, when properly configured, can accurately interpret the *intent* at least seven or eight times out of ten, dramatically reducing the time it takes the user to achieve what he or she needs to, decreasing the number of calls delivered to agents (because the machine will do their jobs), and improving the overall contact center experience (because the agents are freed up to better cope with more complex inquiries where computer systems are incapable of helping).

Language Understanding in Feedback Management

One of the areas in which language understanding systems play a significant role is in **customer feedback management** (CFM). Customer feedback comes in many ways: companies may request it in a form of customer survey, applications, or questionnaires distributed through SMS links, emails, and so on (so it is organized and easy to analyze), but the bulk of it comes unrequested on social media in the form of comments on the company's social media pages, such as Facebook or Instagram. The challenges faced with such feedback are that, first, it is completely unorganized in any way, and second, it often comes in an overwhelming amount. Third, it can often have a huge (and usually negative) impact on the company's business if not quickly responded to.

Nowadays, language systems use their artificial intelligence and language models to help companies effectively manage the above challenges. First, they can easily recognize the topic of the comment. Is the customer talking about the bad service he just received in a branch of the company or the product in general (and which of the products?), or maybe the comment is something more general about the brand. Second, the sentiment of the comment can be recognized—is it a bad comment or a good one? Maybe it is a question the customer is asking, for example, if the new car a company has just released comes with automatic or manual transmission. Third (this is not specifically a function of the language understanding system but an important function of such systems in general), we can measure the impact of the feedback on the business. Feedback from a public figure with thousands of followers will be visible to many people and typically should have a higher weight prioritized than a post from someone with ten or 15 followers on Instagram. The output of the analysis we have described above (topic, sentiment, weight) is used to prioritize the feedback to decide how to act upon it, which goes beyond the scope of this book. Here, we are simply illustrating how language understanding systems can help to manage large volumes of data efficiently and help companies serve their customers better.

Speech Analytics and Sentiment Analysis in Quality Management

In the previous chapter, we described the process of "quality management," in which calls are evaluated to match the contact center's quality criteria; in other words, to ensure the agents say and do the rights things at the right time. To evaluate the calls, a special person (quality manager) listening to random calls (to a certain extent) to evaluate. When we say "random," this does not necessarily mean that the choice of calls to be evaluated is made in a completely random manner—there are rules to ensure certain aspects are proper attention. For example, the plan may be to evaluate ten calls a month for every agent, with five of them selected randomly in which the agent answers support-based calls and the rest in which the agent answers a sales call. This is how quality can be measured to reflect different skills, different teams, and different services, and this approach gives a more detailed picture of where performance is good and where improvement is needed. The problem of the above approach is that it does not scale well, as issues may be hidden in large volumes of calls. For example, an agent may make a serious mistake in only a few calls, but if only five percent of his calls are evaluated, there is a good chance the issue will be missed by the evaluator, at least for some time. To sort it out, we need to evaluate a larger percentage of calls. In the ideal case, this would be all calls, but this is not possible without using some sort of automation because, for example, a contact center receiving 10,000 calls a day can only have two or three quality managers able to evaluate only 200 to 300 calls daily at the most.

The speech and language technologies in this situation becoming lifesaving. At first, we can use speech analytics (SA) technology, which is capable of recognizing certain patterns in recorded calls, some of them are language-dependent, some are not. Examples of language-independent events that SAs can recognize include long, silent periods in the calls (typically meaning that the call has been put on hold, which can indicate dozens of different issues), crosstalk (in which the agent and customer talk at the same time which could mean there is an argument), or a high-pitched voice (typically meaning a hot argument has ensued). Language-dependent features mean that a system can automatically recognize keywords or phrases in recorded calls. With speech analytics, call centers can start quality management in an automated or completely automatic way. **Automated** means that the call evaluation itself is still done by the quality managers, but speech analytics help to spot "problematic" calls out of the thousands of recordings in the system. **Automatic** means that the entire evaluation is done (hopefully, accurately) by the computer system without human involvement.

When it comes to language technology, its ability to recognize and classify the topic and sentiment of conversation may help to significantly optimize the quality management process, making it both faster and more accurate, and this is one of the directions in which the industry is currently developing.

Language technology can also help with quality management for digital channels such as chat and email. We obviously don't need speech analytics here, as emails and chats are already stored as text, but sentiment analysis can be rather helpful here, perhaps even more so than with voice calls. This is because it is rather easy to recognize an argument on a voice call by the high-pitched voices and crosstalk, but we don't have that luxury when dealing with emails and chats. Thus, using sentiment analysis to recognize, for example, emails from unhappy customers, can be rather helpful.

Language Technology in Chatbot Applications

After people invented the telephone, it quickly became the primary communication method for customer service, and it kept this position for the entire 20th century. Now, we see things changing. There were a few "candidates" that could, potentially, become more popular channels for communicating with contact centers, such as video calls, web chat, emails, and social messengers (Whatsapp, Facebook Messenger, and many more). People prefer to communicate with companies using the same methods they use to communicate with each other, and this is where social messengers can truly take over as the new kings of communication.

Messengers—such as WhatsApp, Telegram, and Viber—combine the best features of the other channels. They can be used to chat in real-time but also to postpone answering to a later stage. Unlike phone conversations, they can carry media content, such as pictures or sound files. They have multiple ways to express mood easier, such as emojis. Messengers are typically used on smartphones, so people have access to them all the time. Given the above factors, the demand for companies to provide services via social messenger apps is higher than ever, and most companies are either doing it now or plan to do it in the near future.

A customer checking his account balance and a passenger booking the tickets from his mobile phone using Facebook Messenger and chatting with a chatbot. © IST Networks. Used with permission

Now is the time when customer service over messenger channels is transforming the industry in the same way voice transformed it a few decades ago, via adding automated services. We can even have the same concept as an IVR working via WhatsApp, but with a bit more flexibility, as we are no longer limited to our phones' digital keyboards. For example, to check your credit card balance, we might ask the customer to type and send "card balance" via chat, and he or she will get the answer after a few seconds via the same WhatsApp conversation. This is called a directed conversation chatbot, simply because the conversation is directed by the system, and the user is limited to the set of "commands" we offer from the beginning (in the same way we limit his or her IVR choices with options one, two, three, and so on). With the development of AI-powered language understanding systems, it is now relatively easy and affordable to configure chatbots to understand simple commands in human languages. For example, we can type, "I want to check my account balance for account 015920493-014-USD," and the system will understand this request and come back with an answer (assuming this kind of request can be fulfilled by the system in the first place).

Chatbots typically consist of a few modules, including:

- Artificial intelligence engines—responsible for understanding human speech and mapping it to a function the chatbot is able to perform.
- Backend integration—responsible for performing services the users ask the chatbot to perform, for example, if a customer asks to block a credit card, then we need to establish integration with the bank's credit card system to carry out the task.
- Agent escalation—if the chatbot fails to understand the customer's needs or faces any other problem, it may be desirable for a real human to step in and continue the conversation with the customer. The agent escalation feature normally transfers the entire conversation history to the agent so that he or she can quickly understand how far the chatbot got and continue from there.
- Channel interface—where the chatbot integrates with the actual communication channel. For example, if the company decides to integrate the chatbot with WhatsApp, Facebook Messenger, Viber, Telegram, mobile app, web chat, or a similar channel through which it is important to provide services.

The biggest challenge of the chatbot is developing AI engines (or enhancing and fine-tuning them) so the human communication can be accurately understood. The use of messenger and chat applications has created whole slang sub-languages for almost any language with multiple countries or regions, and even city-specific words. Often, words in foreign languages are spelled with Latin letters. Industry-specific terms are often used, and product names relevant to a particular industry or company have to be added to the dictionary. People also tend to make spelling and grammatical mistakes and so on, so companies developing and implementing chatbots have to keep this and the other above specifics in mind if they want their solutions to have high accuracy when interpreting what users are really asking the system to do. Because messenger applications are on the rise, the demand to have them is huge, and this will stimulate the industry to develop better solutions in the coming years. We will definitely see a chatbot revolution in the customer service industry very soon.

Voice Biometrics

Voice Biometrics (VB) is an example of biometrical authentication. To explain this further, we must first define **biometric** and **authentication**. Authentication is the process by which a person proves that he or she is really the one he or she claims to be. For example, if you go to a bank and demand to withdraw a few thousand from your account, the bank will ask you for your name and your account number, and ask you for an ID or passport to prove that you really are that person who owns the account. If the person in front of us claims to be John Smith, how do we know he is, indeed, John Smith? There are multiple ways to do this, but all of them fall under three categories:

- something only John *has* (like a valid passport);
- something only John *knows* (a secret phrase, answer to a secret question, a secret PIN number); and
- something only John *is* (his finger with his unique fingerprint).

The last one is an example of a biometric authentication mechanism. It is known that human fingerprints are unique to every person, and there are no two humans on the planet with completely identical fingerprints. The same applies to the iris—it is unique to every human being. So is the voice. Voice biometric authentication is a way to authenticate people, using unique characteristics of their bodies, in this case, the voice.

Here is the way it usually works: at the beginning of an IVR session, the user is asked to pronounce a typical phrase—such as "At company X, my voice is my password"—and the characteristics of the user's voice are analyzed by the biometric authentication system and compared to the characteristics of the voice stored for the user's account. If they **match**, the user is considered authenticated; if they **mismatch**, the authentication is considered to have failed; if they **partially match**, the user may be asked for some additional information, such as his or her PIN number, or secret question. He may even be sent a one-time password to his registered mobile number just to be sure.

The definition of a *match* is critical to balance the security and usability of the system. If we demand a 99.9999% match, the chances are that only a few people will be able to pass the authentication because matching one's own voice is never completely possible as our voices tend to change from day to day or at different times of the day. If we decrease the match barrier to 98%, then chances are there will be one out of maybe thousands of people with almost similar voice characteristics who will be able to cheat the system and log in. This is the reason that voice authentication is used most often in combination with other authentication methods, especially for critical transactions. For example, a bank can allow users to use voice authentication to login to the IVR to check for balances and the last few transactions and pay bills, but transferring a large amount of money may require the user to pass an additional authentication level using a PIN number or token. This way, the system stays both convenient (a voice is something we have naturally and don't need to remember—such as PIN numbers or passwords—or carry with us, such as tokens) and secure because anything "critical" immediately requires an extra (stronger) level of authentication. This is where fraudsters will be effectively stopped: at the second line of defense.

Voice authentication can be done at the beginning of an IVR session in which the IVR prompts: "Please repeat <at company X, my voice is my password>," or it can be used as an extra method of authentication when a user is already talking to an agent. In this case, the user may not even know he or she has been authenticated, but because he is talking to an agent, the system can monitor and analyze the user's speech in the background and show a "green light" or "warning" to an agent, meaning, "The person you are talking to really seems to be John Smith as his current voice matches the stored voiceprint," or "Take care—this seems to be a fraudster trying to sound like John Smith as his current voice does not match the voiceprint we have on file for John Smith."

All of the above methods need one important prerequisite to work: besides the voice biometric software, we must have the voiceprints for the users, demonstrating the combination of the unique attributes of their voices. This is important to mention, as the voice biometrics do not store the actual voice samples—they only store the **voiceprints**, or measured voice characteristics. This means that before we can use voiceprints to authenticate users, we have to build a database for the voiceprints. The process of **enrolment** registers each known user in the database and creates and stores the corresponding voiceprint. Enrolment can be done in two ways. The first option is to ask the users to call a special number in the IVR or visit a branch. In either case, the users will be asked to pronounce a few key phrases so the system can analyze their speech and store the results as the users' voiceprints, which will later be used to authenticate them. The second way is completely different and transparent for users, but it requires a more complicated voice biometric system, relying on the fact than almost every contact center has a voice recording system, and thus, it literally has a fair amount of recorded voices for any user who has called there at least a few times. These calls can be retrieved, the agent's voice separated from the customer's voice (this is usually already the case with all decent recording systems), and the remaining customer voice samples are analyzed by the voice biometric system engine to build a voiceprint and store it in the database. With proper integration, this process may be done automatically, eliminating the need for customers to enroll manually, making it more convenient.

Summary and Quiz

The speech and language technologies described in this chapter greatly influence the development of the contact center industry. They are developing rapidly. What was only fiction a couple of decades ago has now become a reality. I personally enjoy seeing how this transformation is happening in different countries and is accepted by different cultures and generations.

Please answer the following self-check questions to test your knowledge of the topics covered in this chapter:

1. The contact center in a rapidly expanding retail company has difficulty recording voice prompts for its new products as they are released. It often takes up to a week to get the prompts recorded and back from the studio. Which solution can help the company solve this problem and have the recordings quickly available by the time they are needed?
   a. Online TTS system
   b. IVR system with integrated TTS
   c. Offline TTS software

d. Speech recognition technology

e. Dictation system

2. Which of the following cases is a good use case for an online TTS?

a. Reading the account balance to a customer via IVR in multiple currencies.

b. Announcing the status of flights.

c. Pronouncing the total amount of an unpaid mobile bill.

d. Welcoming customers by name.

e. All of the above.

3. Which of the following is responsible for the rules of a language and how text is pronounced in a TTS system?

a. Language metadata

b. Transcription rules

c. Voice collection

d. Language model

e. Artificial intelligence

4. You are building an IVR in which users can get a weather report for the major cities in your country. Which of the following will be very helpful to avoid a complex menu structure, given that you want to use the simplest solution to avoid high costs?

a. NLU-powered ASR

b. Offline ASR

c. TTS

d. ASR with a sentiment analysis engine

e. Dictionary-based ASR

5. Which processes may benefit from speech analytics and language understanding technologies (choose all that apply)?

a. Feedback management on social media.

b. Building user-friendly IVR systems.

c. Serving people via social messenger platforms.

d. Quality management.

6. What is the main benefit of using automatic quality management powered by speech analytics (choose one)?

a. Biased evaluations are avoided.
b. Evaluation criteria are not required.
c. Larger portions of calls can be evaluated.
d. Evaluations are more accurate than the ones done by humans.
e. Agents get results in real-time.

7. Which three (3) elements should be measured and evaluated when managing feedback on social media?
   a. Topic
   b. Sentiment
   c. Dialect
   d. Weight or impact
   e. The amount of comments from the same person

8. What are the benefits of voice biometric authentication (choose the two (2) best answers)?
   a. Can be done on a phone call.
   b. Does not require information about the user to be present in the system.
   c. Does not depend on something that can be stolen or lost.
   d. Is more secure than other means of authentication.
   e. Completely replaces tokens.

9. What is the biggest challenge associated with chatbots?
   a. Difficult to integrate with social media applications.
   b. AI models still need to be developed and enhanced to understand human languages better.
   c. They lack security mechanisms as social chat systems are unsecure.
   d. Requires agent escalation for complex cases.
   e. All of the above.

10. Which of the following are true about voice biometric authentication systems (choose all that apply)?
    a. A small chunk of voice must be stored for every user (about 15-20 seconds).
    b. All users must enroll in the system by going to a branch or calling the IVR.
    c. Matches, partial matches, and mismatches must be carefully defined by the administrator as they will impact the usability and security of the system.
    d. Users must say a complete key phrase to be authenticated.
    e. A voiceprint is a set of the different characteristics of a person's voice.

Day 6: Measuring Performance and Customer Experience

If we want to be good at something, we have first to define the criteria to decide if we are good or not. Next, we need to know how to control and measure these criteria, or it will be difficult if not impossible, to improve. Customer service and contact centers are no exception to this rule. In this chapter, we talk about measuring the users' experience in the contact center, the parameters we typically measure, how we can measure them, and how can they be improved.

Despite the fact that the above statement may seem obvious, I regularly see cases in which companies try to build the *best* contact centers without understanding what *"best"* means for them. I usually ask the question: "When, 12 months down, the road your system is up and running, the agents are answering calls and emails, and tickets are opened and closed, how will you decide if the project is a success and/or what needs to be improved? And if you think there is something that needs to be made better, how will you decide what that is and how to improve upon it?"

Those who have some answers to the above questions tend to have contact centers that support their business better than the ones who simply think about hiring a certain number of agents and giving them a system with which to answer customer contacts.

To define these criteria, there are two fundamental questions about which we have to think:

1. What do the customers expect to get (why do they call, chat, send emails to)? Do they want to know their account balances? Do they want to block their credit cards? Book flights? Renew insurance? Thinking about this will help us understand what people really need to do, and it will be easier to answer questions about what kind of service they expect to get. For example, when a user wants to block his or her lost or stolen card, the queue waiting time should be minimal because people with lost cards are likely to experience a high level of stress at the time of contacting the call center, so quick serve is of paramount importance.

2. What does the business what to achieve? Do we see our contact center as a department to only support the main business, or is the contact center a department directly involved in generating profit. An example of a "supporting" contact center (we call it a **cost center**) can be one in an electronics shop whose primary role is to answer customer complaints and inquiries, call for a technician, help customers with the first use of their washing machines or microwaves, and the like. An example of a profit-generating contact center (we call it a **profit center**) can be an airline contact center that provides the function of booking tickets, adding extra paid options to tickets such as baggage, paid seat selection, or onboard meals.

Usually, contact centers carry out both functions, acting as cost centers supporting the business, and profit centers, generating a profit by selling goods and services the company offers; therefore, the criteria for these functions (sometimes they are separate teams or divisions) will be different.

Let us now dig deeper into the kinds of metrics and their meanings. For simplicity's sake, they have been divided into groups, which is a conventional division.

Load and Sizing KPIs

These KPIs (KPI stands for **key performance indicator**) show how many contacts (calls, emails, chats, etc.) the contact center handles and how well it copes with the load. For the voice contact center, the primary unit is the **call**, and the KPIs showing the load are:

- **Busy Hour Call Attempts (BHCA)**—shows how many calls the call center receives in its busiest hour (the hour in which the load reaches its maximum). BHCA is important when planning contact center resources such as the number of telecom lines, IVR ports, agents to employ, and so on. BHCA is usually much higher than a normal, daily contact center load. In a retail bank, for example, the BHCA can be reached in the first half of the day on Monday corresponding to the end of the calendar month, simply because the bank customers are trying to check their account balances using the IVR. In telecom companies, the BHCA is often reached on the dates on which customer bills are delivered. This is one of the reasons some telecom companies now have fewer billing dates, on the 1st, 11^{th,} and 21st of each month rather than once a month—to reduce the load by dividing it across several days.

- **Number of contacts handled (per hour, day, month)** —this is the number of contacts (including calls and digital contacts like email, chat, and others) the contact center handles in the specified time interval. This metric is used to size the system properly, the number of agents in the contact center, and perform a high-level load analysis. Drilling down and checking the number of contacts per channel and the way it changes year after year may reveal interesting trends about how fast the customers' channel preferences change from voice to digital channels.

- **Average talk time (ATT)**—the agent-related metric showing how much time an agent spends on average, talking to each customer on the phone before the call is closed.

- **Average handle time (AHT)**—similar to above but also includes the after-call work (ACW)—also known as **wrap-up time**— required after a call has been terminated. For example, an agent may use this time to record post-call feedback or write a follow-up note in the CRM system, send an email, or simply take a deep breath before the next call. This means that AHT = ATT + ACW.

- **Average queue waiting time**—the average time a customer has to wait in a contact center queue before the call (or chat) is transferred to a live agent. The queue time is counted from the second the customer chooses the IVR option to talk to an agent to the second the agent picks up the call to welcome the caller.

- **Service level**—the characteristic showing how fast agents respond to calls, emails, and chats. There are two formats for showing service level, and I will explain the more complex of the two here: the two-digit service level format, recorded as XX/YY. YY is the time it should take between the customer choosing the "talk-to-an-agent" option and the time the call actually gets answered by an agent. XX is the percentage of calls that have been answered within the YY time period. For example, if a service level is 80/20, it means that 80% of the calls are answered (or should be answered) by agents within 20 seconds after the customers' request. There is a difference between **service level target** (what a contact center wants to achieve) and the **actual service level** (what a contact center really achieves). Actual service levels can be higher or lower. Some contact centers prefer to indicate service levels in a single-digit format. In this case, they might say that their service level is 15 seconds, and their service level achievement rate is 94%, meaning that they manage to have 94% of the calls answered by agents within 15 seconds after the customers' requests.

- **Abandoned calls ratio**—the percentage of calls in which customers hang up while waiting in the queue, calculated as a percentage of the number of calls in which customers *request to be connected to agents*. Remember that the request to talk to an agent does not happen with every call, as in some cases, the customer may just use IVR services and hang up without requesting to talk to an agent. Such calls will not affect the abandoned calls ratio. Abandoned calls are usually further classified into abandoned within service levels ("good" abandoned calls) and after service levels ("bad" abandoned calls). For example, if we receive 1,000 calls out of which 500 request the agent transfer option, and ten of those leave in the first few seconds after requesting it, and ten more leave after waiting a good few minutes, we then have a total abandoned call rate of four percent (which is 20/500), whereas two percent are abandoned outside of the service level (10/500) and another two percent are abandoned within the service level (10/500). Some algorithms ignore the "good" abandoned calls in their calculations or offer a configurable threshold (usually a fraction of service level goals) to exclude these calls from calculation. For example, if a service level is 15 seconds, any calls abandoned in less than 5 seconds are neither considered for service levels nor are they accounted for in abandoned call calculations.

As you may already have guessed, the above KPIs are tightly related to each other. If we increase the service level goal, the ratio of abandoned calls within the service level grows. If the load is higher, queue times increase unless we also increase the number of agents. Examples of service level goals from the industry are usually between 15 and 30 seconds, with the percentage goal being between 80 to 90%. The most common values are 80/20 or 85/20, except in cases related to emergencies or high priority services, where service levels are usually higher. For example, an emergency contact center—such as 911—usually offers service levels close to 100% within less than a couple of seconds—almost immediately—as it should be.

Cost per Contact

Cost per contact is a fundamental metric representing what the company pays on average to handle a single customer contact. If we are talking about a global cost per contact (covering the entire contact center operation), then it is the sum of all operating costs divided by the number of contacts the contact center handles. When we talk about operating costs, we mean the sum of capital expenditure (CAPEX)—buying and configuring contact center technology solutions, including equipment and software licenses, furniture, agent PCs, and networks, for example)—and operation costs (OPEX), such as the cost of telecom lines, equipment maintenance costs, software subscription or support costs, agent salaries, and so on. Because the CAPEX doesn't typically happen every year (if we build a contact center solution it is normally in working order for three to five years, until the next major upgrade cycle, considered another CAPEX), we should consider this, and calculate the cost per contact over one such period instead of only using part of it. OPEX costs happen regularly and are related to regular maintenance of equipment and software, technical support, agent costs (salaries and benefits), facility costs, telecom costs (phone lines and Internet connections), and any other expenses the contact center division has *on a regular basis.*

Cost per contact can be counted for the entire contact center or it can be for different divisions (for example, in a bank, the general retail and gold customer divisions may choose to calculate their cost per contact separately). Cost per call, email, or chat are also good examples of cost per contact calculations. It may be interesting to compare cost per contact for different kinds of industries (such as banking, telecom, or retail), different kinds of contacts (such as agent calls, IVR calls, chatbot sessions, or email) and even in different countries. Then, comparing and analyzing these numbers can reveal many useful insights about the direction in which the industry is developing. These numbers may also explain the background behind some business decisions that seem rather surprising but will start to seem completely logical if we simply "see the numbers" behind the reasoning.

Let us take a few moments to discuss some variables that directly affect the cost per contact metric. Usually, companies try to maintain reasonable costs per contact, which are not much higher than the industry standard. To achieve this, they seek ways to serve more contacts and reduce costs at the same time. If you remember the discussion we had in the chapter on WFOs, more than half (often up to 70%) of the operating costs are related to contact center agents; therefore, the typical and correct approach is to invest in a better technology because the reduction of agent-related OPEX will quickly cover and overcome the investment in technology.

If you have ten workers digging holes, you might think of a few ways to increase their performance; you could reduce the duration of their breaks; you could ask them to dig faster; you could fire them and hire stronger workers, and so on. A smarter approach would be to provide workers with better spades, or if possible, an excavator. While this is a logical approach, it is often overlooked in the contact center industry. I have seen cases where, instead of investing in the proper agent tools, the management hires more and more agents to cope with the increasing amount of contacts they received. Hiring extra agents creates additional, agent- and technology-related costs. Every agent needs the equipment to work (such as PCs, phones, and contact center systems). Agents are also getting their salaries every month. In addition to the above costs, the agents need someone to manage them, and increasing the number of agents in a contact center also means hiring more team leaders, supervisors, quality managers, and other support staff (with their own associated costs). I have personally seen call centers in which hundreds of agents spend time switching between multiple applications, manually entering customer data, and asking customers unnecessary questions to get information that should already be available from an IVR session. With an increase in the number of contacts, service levels begin to fall and supervisors and team leaders apply additional pressure on their teams to work faster, resulting in poorer quality.

The smarter approach would be to invest in a technology such as unified agent desktops, with which agents can see all the information they need on a single screen. This includes information about the caller as well as data from the IVR session, which automatically displayed to the agents so they don't need to ask extra questions, and they can finish the conversations quicker. With this approach, contact center agents can handle calls faster, average handle time (AHT) will be reduced, and first call resolution (FCR) will increase.

Technology can be a part of the solution to the problem of reducing the average cost per contact. Labor costs are much lower in some countries, and it often makes sense to have contact center divisions located in countries where agent-related costs are reduced. This approach is an example of a global trend toward moving some labor-intensive activities overseas. It is also the reason why we may be ordering our pizzas from a restaurant just around the corner while the phone is being answered by someone located on the opposite side of the globe. It still sounds a bit unexpected when "around the corner" guys ask you for your exact delivery address starting with the city (come on—you should be able to see my building from your window), but it makes sense when you consider that you may be talking to lady sitting in a large call center somewhere in the Ukraine. Once you tell the lady that you need two large, pepperoni pizzas and a salad, she finalizes your order in the system, clicks "submit," and the order appears a few thousand kilometers away on a pizza machine, and is delivered to your house by a delivery boy within 25 minutes. This finalizes the story of your lunch having literally traveled around the globe in less than 25 minutes to reach your house. What is even more astonishing is that this is the most cost-effective and convenient way to do it.

Optimizing (we say *optimizing* not *minimizing*) the cost per contact can easily be the topic of a separate book, and I don't want to go into a deep discussion here. Instead, I will summarize a few points that are important to keep in mind:

- A good way to reduce cost per contact is to use technology wisely rather than push the agents to talk faster or reduce rests between calls.

- Because agent costs represent the majority of the costs, the usage of self-service channels—such as IVR and chatbots—is a relatively easy way to reduce cost per contact.
- Some requests can be served better over specific channels, thus having channel diversity can help to maintain a low cost per contact.
- Cost per contact must never be considered without other metrics, such as FCR. It is better for a business to have one long call of five minutes and solve the customer's issue rather than two calls lasting three minutes each. Though the cost per contact is lower in the second case, the overall cost of solving customers' issues will actually be higher.
- Cost (overall or per contact) should not be the ultimate focus of contact center operation, even if the contact center only acts as a cost center. Managers should always remember that the *ultimate goal of the contact center division in the company is to contribute to the company's success in the market.* It is often fine to have a higher cost per contact if it builds customers' trust and makes them more loyal to the company. Such a company would build a more stable and profitable business because *customers don't look at the contact center KPIs. Instead, they look at how well they are served and if their problems are quickly resolved.*

Customer Experience KPIs

In the previous section, we said that it is ultimately more important for a company's success to provide high-quality service to its customers rather than optimize internal operational KPIs. As a customer, you don't really know XYZ's call center's KPI, but you know that when you had a problem with their product, they answered your call and solved your problem quickly and professionally. When you lost your user's manual, they quickly sent you a copy of the same via WhatsApp, free of charge. This is what makes you recommend XYZ to your friends and colleagues. It is also what makes you go back to the same store if you need anything else. Your **experience** with them is great, and you become their **promoter**.

It is obvious that all companies would love their customers to experience a story similar to the one above, but the challenge is—as usual—finding a way to measure the **customer experience**, something we must do in order to improve upon it.

There are a few metrics to measure the experience, as follows:

- **First call resolution (FCR) or first contact resolution if the contact center is multichannel**—the ability to provide what a customer wants in the first contact (i.e., without the need for a customer to call or contact us more than once). FCR is easy to define, but it may be rather tricky to measure. If a contact center receives thousands of contacts daily, how can we know the percentage of contacts in which the customer does not have to come back to us twice? There have been multiple ways suggested to measure this, including the wrap-up (in which agents fill-in the FCR's yes/no field in the after-call work if the call was, indeed, finished), or running a complicated report combining information from the contact center system and the ticketing—or CRM—system. We can also measure the number of repeated calls, assuming that if the problem is not solved, the customer will call us back again shortly. All of the above methods are inaccurate. For example, if I call an airline in the morning to book my ticket and do so successfully, but in the evening, I call again to buy extra baggage, this might appear as a repeated call in the system, though a perfect FCR was achieved in both cases. Or an agent may mark the problem as solved (and he truly believes it has been), but a few days later, the customer has the same problem (sounds familiar, doesn't it?). The "internal" method of FCR measurement is often subjective or inaccurate. An "external" method of measurement may be a better alternative. After a call has been finished or an email or chat has been answered, the customer may be asked to answer a question, such as, "Did we manage to solve your issue on the first try or did you have to contact us several times?" Although not every customer will choose to answer the survey and the customer's perception about the problem being solved is not always accurate, this method has the advantage of receiving important feedback from the person we are ultimately trying to satisfy: the customer. We can arguably say that measuring the customer's "perception" of the FCR is the only correct way because the entire customer experience is about how they perceive the service and not about the numbers we get in our reporting systems.

- **Customer satisfaction score (CSAT)**—a metric showing how customers are satisfied with the service they receive. The usual ways to measure customer satisfaction include a post-call survey on the IVR or a survey after a chat session in which customers are asked if they were satisfied with the service provided. Good customer satisfaction is not equal to having loyal customers—you might pay for an airline ticket, have a decent trip, receive your baggage quickly on arrival, and generally feel satisfied, but this does not necessarily mean that you will recommend the airline to your friends. On the other hand, you may experience a flight delay or cancellation, but the airline managed it well by providing you with an alternative booking, hotel reservation, or compensation in the form of miles or free tickets, and you'll still be satisfied. You will also probably share the **experience** with your friends about the way in which that particular airline fixed the issues to make you happy. This is the reason why, instead of measuring CSAT, companies choose to use the **net promoter score (NPS)**

- **Net promoter score (NPS)** is the metric showing how likely the customers are to recommend the products or services they received. NPS is usually measured on a scale of one to ten, where marks of seven and eight are considered neutral. Anything less than seven represents an area that needs improvement, and nine to ten usually means that the customer was not only satisfied but ready to promote the product or brand. NPS is measured in the same way as CSAT, via surveys, questionnaires, and feedback forms. We can arguably say that a properly measured NPS makes CSAT and FCR measurements excessive and often unnecessary. An NPS is more effective than CSAT and more accurate than FCR, and in the end, it is easier to get customer feedback when there is only one question in the survey rather than a few (yes, customers are usually lazy to answer surveys despite being convinced that their feedback will help us to serve them better).

Having discussed multiple KPIs, it is time to talk a little more about the tools we can use to measure these KPIs.

Monitoring and Analysis Tools

Monitoring of contact center KPIs is done using a multitude of tools, both real-time and historical. Real-time means the data is no older than 15-30 seconds—most systems can display data as fresh as five-seconds-old, and this is more than enough, as operating decisions in contact centers usually take a minimum of several minutes to be made. Historical reporting means looking at "old" data, where old may be defined as one hour ago or a period of a few months or even a few years.

Data used for all kinds of monitoring and reporting is collected in multiple ways. For example, the ACD system stores records about each interaction it processes with a great level of detail, and these data are stored in the historical reporting database. The IVR system may have its own data store for data, including information about each call as it moves through the IVR system, chosen menus, time it took customers to use the services, and so on. While telephony, customer feedback, ACD, IVR, customer feedback, and some other systems are the primary sources of information measurement, display and analysis is also done using the following modules:

Wallboard—a large screen that can be easily seen by the teams, displaying operational real-time or recent information to agents and their supervisors. A well-designed wallboard software uses the screen wisely, displaying information in an easily readable format, often in a way that displays a lot of parameters a single screen without clutter. The customer service operation is often affected by external factors (for example, the situation in the bank's contact center may be affected by stock market volatility, and the situation in the airport's contact center will definitely be affected by bad weather conditions). This is why a good wallboard software can display information from external sources, such as stock markets, weather forecasts, live video streams, currency fluctuations, and any other external data.

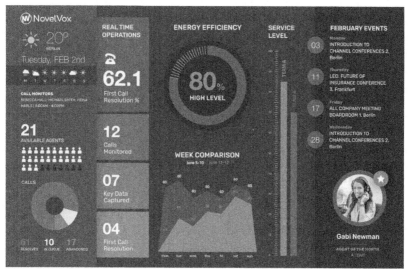

Novelvox iVision Wallboard showing real-time contact center statistics on the screen.

Dashboard is another system designed to display real-time or recent information on the supervisor's screen. Most dashboard applications today are web-based, meaning they are accessed using a web browser from any computer or mobile device. Dashboard applications normally offer a good level of personalization so that supervisors can design their own layouts of the dashboard to monitor the KPIs he needs the most. Good dashboard applications normally support external data sources so the online weather forecast or stock exchange data—and even the number of currently opened support tickets extracted from helpdesk system—can be displayed next to the current number of calls in the queue and the current service level, measured over the last few minutes.

Dashboard and wallboard applications are similar as both read and display recent data in real-time in an easily readable format. Many people even use the terms wallboard and dashboard interchangeably, but I submit that this is not entirely accurate, as the wallboard software's design is optimized for being displayed on large screens, typically hanging on the wall (as the name suggests), while dashboards work on computer displays and are typically personalized for contact center supervisors and managers.

Besides real-time data displays, contact centers use historical reporting tools for building reports using data aggregated over a period of time. An example of a historical report can be a service level achievement percentage, measured in averages of 30-minute intervals over the past 30 days.

Historical reporting systems can use web-based interfaces or have dedicated applications that need to be installed on the computer. Contact center vendors typically provide some pre-built reports with which to start, and they also offer good capabilities for modifying existing reports, including adding or removing fields, changing time periods, building reports in table or graph format, and so on. They also have export capabilities with which to export data in spreadsheet format (Excel or CSV) for further analysis.

WFO applications—which were discussed in the previous chapter—also offer very rich and informative reporting and analysis tools. This is especially true of workforce management systems (WFM), with their core function of forecasting and scheduling which relies on deep data analysis and prediction algorithms.

Customer feedback tools often have their own reporting modules to display and analyze data.

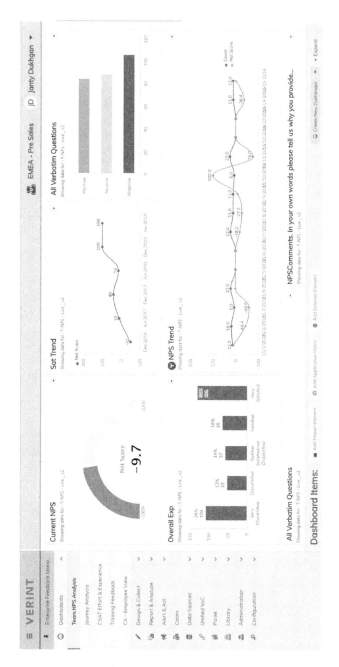

Verint Enterprise Feedback Management system dashboard showing the customer experience KPI's. © 2020 Verint Systems Inc. All Rights Reserved Worldwide. Used with permission.

Nowadays, companies seem to prefer aggregating historical data from multiple sources into one data store so it can be analyzed using advanced mechanisms, such as **big data**. This is why ease of integration (more specifically, ease of extracting information) is perhaps one of the most important characteristics of any system or module in the contact center infrastructure.

Summary and Quiz

In this chapter, we learned how contact center performance and quality can be measured and which KPIs should be measured and why. We also discussed the tools to measure and display KPIs.

Please answer these self-check questions to test your knowledge of the topics covered in this chapter:

1. Which metric shows the maximum number of contacts the contact center receives per hour on the busiest time of the busiest day?
   a. Service level goal
   b. Summary load per hour
   c. BHCA
   d. Number of contacts handled per hour

2. With an ATT of 44 seconds and an AHT of 66 seconds, which of the following is true?
   a. The service level is 66.6%.
   b. The ACW is 33% of the AHT.
   c. The potential efficiency gain is 22 seconds per call.
   d. The Service level is 33.3%.
   e. None of the above.

3. The BHCA is 10,000 calls, while the number of calls answered by agents during the busiest hour is 7,000. Which of the following statements is the most accurate?
   a. 3,000 calls were abandoned.
   b. The service level cannot exceed 70%.
   c. The service level is not less than 70%.

d. Such a situation cannot happen in real life; therefore, there is probably something wrong with the reporting system configuration.
e. None of the above.

4. A contact center receives 100,000 calls within one hour. The target service level is 75/15. According to the IVR report, only 80,000 customers requested to be transferred to a live agent. You check the agent report and discover that 78,000 calls were answered by the agents, 1,500 calls were abandoned after waiting for more than 15 seconds in the queue, and 500 were abandoned during the first 15 seconds. Which of the following statements is the most accurate?
   a. There is not enough information to measure the actual service level.
   b. The actual service level exceeds the target.
   c. The actual service level is below the target.
   d. The actual service level is on target.

5. Which two (2) of the following are the best ways to optimize the average cost per contact?
   a. Try to measure the FCR rate analysis and work to increase it.
   b. Try to measure the FCR rate analysis and work to decrease it.
   c. Decrease the ACW.
   d. Increase the ACW.
   e. Reduce the AHT by designing and implementing an integrated agent desktop CTI.
   f. Increase the AHT by designing and implementing an integrated agent desktop CTI.

6. You are the contact center manager at a large retail chain selling house appliances and consumer electronics. The contact center supports customers who have difficulties using the products they have purchased. The contact center also promotes the customers' purchase of accessories and ordering regular maintenance services for the products they already have. The company's CEO asks you to optimize operation costs for the contact center. Please specify

which of the following would be the best ways to deal with the request (choose all that apply):

    a. Automatically implement a smart agent desktop CTI, showing the customer's name, list of products the customer has purchased, and any open tickets.

    b. Implement a function in the IVR to inform the customer about the status of the maintenance request, if any.

    c. Implement a chatbot that automatically suggests answers to the most frequently asked questions.

    d. Implement an omnichannel contact center so customers can also send requests via chat and email as these channels allow the attachment of data such as pictures.

7. You are a contact center manager at a quickly growing company. The company's CEO asks you to measure how the contact center will support the company's growth. At which of the following KPIs will you look?

    a. NPS

    b. FCR

    c. CSAT

    d. AHT

    e. BHCA

8. You are managing an internal contact center offering IT helpdesk services for a large enterprise. Typical requests include resetting user account passwords and supporting desktops, printers, and office applications. As a part of the overall company transformation and cost optimization program, the CEO asks you to find ways to optimize and improve the services your department offers. Which of the following will you implement (choose the three (3) best answers)?

    a. Measure the AHT and focus on ways to reduce it.

    b. Measure the CSAT and focus ways to increase it.

    c. Implement a chatbot for resetting user passwords.

    d. Implement an IVR system to welcome callers before transferring calls to the agents.

    e. Measure the NPS and find ways to increase it.

    f. Implement support via chat and email to offer users more ways to reach the helpdesk.

g. Hire a technical consultant to configure the system to bring the actual service level closer to the target service level.

9. You have performed a series of improvements in contact center processes over a period of six months. You would like to analyze how this optimization translates into improvements in the contact center's basic KPIs. Which of the following tools will you use?
   a. Dashboard
   b. Historical dashboard
   c. Historical reporting
   d. Wallboard
   e. Big data analysis

10. Which three (3) of the following KPIs represent the best ways to measure customer experience?
   a. CSAT
   b. ATT
   c. AHT
   d. Actual service level
   e. BHCA
   f. FCR
   g. ACW
   h. NPS
   i. Target service level

Day 7: Contact Centers Today

Information and telecommunication technology is developing rapidly, and over the last couple of decades, it has affected all aspects of people's lives like never before. This phenomenon is referred to as "**digital transformation.**" There are many definitions and explanations of this term, but the one I prefer is that it means doing the same things we used to do many years ago—buying coffee, commuting to work, traveling to a different city, paying bills, or getting medical treatment—but we now do these things *differently* due to digital technology. It means technology has changed (or transformed) the way we do things. Contact centers (and customer service in general) have been largely affected by the digital transformation in the way calls are answered, storing, processing, and analyzing the information, the types of communication we can receive in contact centers, and the way we manage this communication. All of it has significantly changed and continues to do so today.

The examples of digital transformation in the contact center that I would like to mention here are:

- **Digital channels**—which, in contact center terminology, represent non-voice channels such as chat, email, SMS, messengers, and so on.
- **Robotic Process Automation (RPA)**—an easy and effective way to automate many processes and not just in the contact center.
- **Office workload distribution**—using the concepts of contact center skill-based routing but applying them to a company's internal organizational tasks.

We already discussed digital channels in the previous chapter, so let's talk a bit about RPA and workload distribution in the sections below. After that, we will talk about another type of transformation affecting contact centers: **cloud transformation**.

Robotic Process Automation

Robotic Process Automation (RPA) is a young, promising technology that automates simple, repetitive, computer tasks in the workplace. This kind of automation is carried out on a normal computer interface in exactly the same way as a human would normally do it, but instead, the software application—called a **software robot**—carries out the entire process faster.

RPAs can be illustrated using a few examples. Imagine that you receive a few thousand emails on a daily basis, and you have to open each one, check if it contains an attachment. If it does, you must save the attachment in a separate folder and name it containing the user ID of the person from whom you've received it (replacing the @ symbol with the word "at") and the date when the email was received. For example, if you received an email from mhilmy123@gmail.com with an attached picture called mypassport.jpg at 11:05 a.m., then you will save this picture in a folder named "mhilmy123-at-gmail.com-19OCT2019-1105AM." The task is easy when you receive only a few such emails a day, but it can take all of your time if you receive hundreds of such emails on a daily basis. A robot might come in handy here, as it can process the entire operation of opening an email, saving an attachment, and renaming it to the correct standards in a matter of seconds.

Another example can be filling out an online registration form, which is not a complex job unless you need to do it for a good few thousand people with all of their details stored in an Excel table. Completing this process one by one, copying names and contact details, submitting the form, and repeating it is tedious and time-consuming. Using a software robot instead, "teaching it" to perform simple copy-paste-submit operations can get the day's job done in a matter of minutes.

Another good example is the case in which a secretary has to spend an hour visiting the websites of the leading banks to get up-to-date currency exchange rates every morning, putting the information into an Excel table, formatting it in a standardized way, and sending it to her manager by 10:00 a.m. every morning.

RPAs help with automating tasks and jobs similar to the ones described above, those that could be nominated for "the most boring and stupid repetitive job in the world" award. To automate these tasks, the RPA uses a robot—a computer program able to emulate simple human actions—to open applications such as Excel or Outlook, clicking certain buttons, switching, typing, copying and pasting, opening websites, finding information, and so on. RPAs are a relatively simple (and thus, inexpensive) but extremely promising technology. In the 20th century, computers helped humans to do things quicker. Now, in the 21st century with RPAs, we can basically teach computers to use other computers to do things even faster. Humans are responsible for the more intelligent parts of the process, such as teaching the robots to do things and controlling them, so they do things properly.

Because it takes time to "teach" the robot to do things, the RPA is only applicable to *repetitive* processes. If something can be done manually in 15 minutes and is only required to be done once, then there is no need to spend a few hours programming a robot to do the action. If, however, this 15-minute-long process is to be done every day, the long-term time savings are more obvious, and bringing in a robot to do a task instead of a human could be considered a good option.

A second important condition for RPA applicability is that the task needs to be **strictly formalized**. This means it has to follow a well-defined procedure with simply defined, step-by-step actions, and should not involve any complex decision making. The description of **strictly formalized** is a politically correct definition of the term "stupid process" I used above, though most people would probably still use the less politically correct term to describe what they often have to do at the office.

RPAs will not work for processes (or parts of processes) requiring human intelligence and complex decision making. For example, companies hire and fire thousands of employees every year, but this is not something an RPA can do—even with well-defined HR procedures, hiring or firing an employee is a complex decision that must be done by an expert or even a group of experts, so an RPA is of no use here. We can still use the RPA to go through a list of the thousands of emails we receive from potential candidates, extract their attached CVs, and upload them into the corporate HR database. This will save the HR people time to complete more complex and intelligent jobs. Once the new hire has been chosen and hired, we can use an RPA to open and configure his or her account in HR, corporate email, the internal social portal, the company's knowledgebase, apply for company insurance and health club membership, and do a few dozen other actions to get the employee on-board quicker. This often makes sense for big companies hiring tens of people every day as the time for onboarding an employee can be reduced from several hours to only a few minutes.

The above example brings us to the concept of human-robot collaboration. What if we had a long process in which most of the steps can be automated, but a few key decisions must be made by humans? This is the case where **attended** robots are used. These robots are supervised by humans who control key decisions and the accuracy of the entire process, while robots do the hard work, performing repetitive, uninteresting tasks that are boring and tend to exhaust people.

In the contact center and customer service domain, RPAs have huge potential because contact centers naturally deal with thousands of contacts (meeting the first RPA condition: repetitive actions), and these contacts often carry similar requests. For example, if you lose your credit card and call the contact center to have it blocked, the agent typically needs to open the credit card management system, paste your account information into the window, make a few steps to block the card, open another window, and paste your information there in order to issue a replacement card, and so on. Some agents can do this in 30-40 seconds; some experienced ones may finish it in 20 seconds; the robot can do the whole thing in less than three (where most of the time will be spent by an agent looking at screens to confirm the robot's actions).

Contact center agents may have the tools and authority to solve many customer issues and respond to basic customer inquiries, but very often, agents are only responsible for taking the necessary information from customers and filing requests in internal systems to be handled by back-office teams. For example, if you apply for a mortgage, the decision to approve it will likely be made by the loans department after analyzing your credit history and requesting loan conditions and other data. Therefore, the back-office also deals with repetitive tasks. An RPA is equally applicable in back-office task automation, which positively impacts overall customer service. In the above example, having a great contact center will not be helpful if it takes 20 days to approve a mortgage application. RPAs are one way to ensure back-office employees are productive on individual tasks. In the next section, you will see how to optimize the overall task management process in the back-office.

Workload Distribution Systems

The contact center is a division that deals with customers directly (just like a branch), but customers are really served by the entire company. If you apply for a mortgage loan, the efficient, polite employee talking to you (in person or via a contact center) is an important prerequisite for good customer experience, but the actual application is not reviewed by the agent. Instead, it goes to the department responsible for loans at the bank, and the experiences you will have with the bank in the future depends on that department. A well-organized and efficient back-office is an important prerequisite for good customer service. It is a typical misconception to think that only customer service or contact center departments are responsible for serving customers; indeed, it is the whole organization that serves the customers. In the previous chapters, we discussed how contact centers are built, organized, controlled, and how to make them work faster and better. We demonstrated how agents can take calls quickly, deconstruct customer inquiries, record and respond properly, and so on, but what if we were to file a request for a car loan through an excellent contact center only to realize that, two weeks down the road, the request is still being processed. This indicates that even if the contact center is doing an excellent job, the rest of the company lacks the same level of performance. *Can we apply the methods and practices of contact centers to reach the same level of optimization in the back-office?*

Let us go back to an earlier chapter of this book in which we discussed methods for distributing multiple calls to a group of people. Hunt groups, ring groups, and different methods of distribution—we introduced them as basic ways to distribute the calls, but then we almost immediately moved to skills-based routing and then into skills-based routing with prioritization as it generates better and more predictable results.

If we look at back-office operation today, we will see that tasks (such as your loan request in the example above) are still often distributed in the same, primitive way. Each employee has his or her own task list and decides which task must be done first, which task is more or less important, what to prioritize, what can wait, and so on. The idea behind workload distribution products is to *treat each task the same way as the call centers treat a call: as an item with a given priority, which needs to be answered within a certain time by a person with the right skills.* The employees in the back office are similar to the contact center agents in that they have skills configured in the system, corresponding to the type of work they can do. They also have software to control their states: ready, not ready, busy with some task, or logged out completely.

This intelligent system (let's call it a workload distribution system, or WDS) monitors all tasks in all systems in which the tasks are generated. Loan requests at the bank, surgery approvals in a medical insurance company, helpdesk tickets at an IT company, a new line activation in a telco company—all are examples of customer requests needing specific skills to be completed with their associated SLAs. The same system also monitors the people available to do the task; it knows that Jim and Jake both have the right skills to approve home loans. The workload distribution system has the capability to prioritize tasks and put them in front of the other tasks, if necessary. A good WDS can even dynamically change the priority of a task to make it more urgent as the task gets closer to its SLA.

The benefits companies achieve by using a WDS are typically significant and financially measurable. First of all, we can ensure that all tasks are fulfilled on time, or at the very least, we can control the process to see how badly the SLA is missed. Second, we can ensure that the tasks are prioritized according to their *importance to the company,* rather than to the employees' *personal choices.*

The last but probably most important outcome is that management will have insights about what is happening. If without the WDS, the only thing they could say is that a typical house loan request takes four working days to be approved, now they can say that in those four days, there is an average of three hours to fill out the required forms, three working hours to receive the customer's credit history, five hours more to analyze it, eight hours to do a property value assessment, and so on. Management can also see which of their employees spent time working on which requests, and they can assess the overall process, and so on. It now becomes possible to recognize and fix issues, make processes faster, remove bureaucratic barriers, and even make fair, unbiased staffing decisions, promoting hard workers, warning those who perform poorly, and so on.

The important thing about a WDS is that it *does not need to replace existing systems* used by employees to do their jobs. Customer data will still be in the CRM, helpdesk tickets will still stay in the helpdesk system, and insurance policies will remain in the insurance system. The WDS simply integrates these systems to sense and understand the tasks created and closed there to manage them efficiently, using the available workforce.

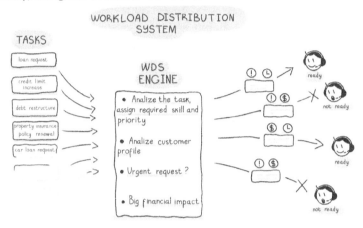

Cloud Transformation

Along with digital transformation, there is another process taking place in the technology world, called "cloud transformation." We can give this the definition we gave to digital transformation, but it is a bit more technology oriented. Cloud transformation means using the same systems—such as applications, data storage, messaging, and analysis—but instead of these applications being installed on our computers (or the servers in our offices), we now have these located in the "cloud," which most often means to consumers, "somewhere on the Internet."

In general, the Cloud approach has advantages and disadvantages beyond the scope of this book. Here, we discuss how the cloud approach is applicable to the contact center setup and the pros and cons. A cloud-based contact center has some or all of its parts installed on the cloud, leaving little to nothing to install on the customer's end. The typical setup will have the contact center's software engine running on the cloud while customer-relevant data is still hosted on the premises. The integration with the phone company may also be done on the premises or from the cloud (this would be easier if the cloud provider and the telco are the same company). Agents can be located on the premises, or they can be located somewhere else on the Internet. The beauty of a cloud solution is that any resources are available for use as long as computers are connected to the network.

The structural diagrams of cloud-based contact centers are shown in the picture below, though different vendors and solutions may have completely different layouts. Three separate options are shown in the pictures, with the phone number either being terminated at the customer's site using
CPE (customer premises equipment), terminated somewhere in the cloud or being provided by a cloud contact center solutions provider.

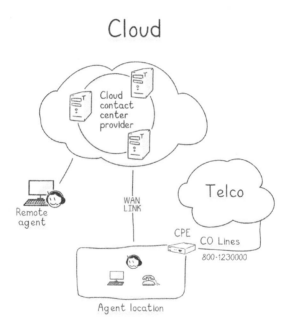

Cloud

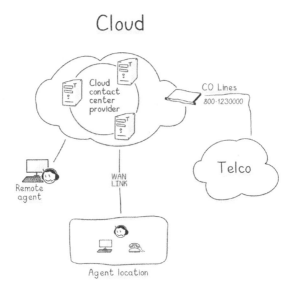

Cloud

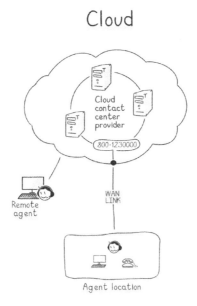

Let us start by looking at the drivers, the issues faced by contact centers today, and how these issues are solved by cloud-based deployments. Then we will try to figure out where the cloud approach is applicable and where it is still questionable or too early to consider.

One natural aspect of contact centers today is their complexity and the high costs associated with them. The complexity comes from the need to serve customers through multiple channels, and very often, new channels have to be added quickly. Features and functions are also rapidly evolving in the industry, and companies feel the need to upgrade contact center software at least as often as a typical consumer changes his mobile phone, which is every 18-24 months. The problem is that an upgrade project for a complex solution—such as contact center—typically takes six to nine months to complete (in a best-case scenario). So, very often, by the time the upgrade is finished and the system has moved into a production state, it is already half-way outdated. We can call this the **challenge of upgrades**. In the case of a cloud solution, the burden of updating, upgrading, patching, and modifying the system to remain up to date is addressed by the cloud contact center solution's provider, and companies don't need to worry about anything except for learning and utilizing the new features as soon as they become available.

The cloud is helpful in dealing with the second challenge: the **challenge of scale**. Very often, companies need to increase the number of agents they have as the business grows. With a premise-based system, this means that new licenses and hardware must be procured and installed, a time consuming and expensive endeavor, but the situation is even worse if the required growth is temporary, for example, if it is seasonal or related to a campaign. In this case, the company may require an increase in the size of the contact center for only a few months before returning to its original size. With a cloud-based solution, this issue is easy to solve with extra resources and licenses that can be allocated quickly and used only for as long as they are required.

Cloud contact centers can offer competitive pricing compared to on-premises systems, especially when you consider that companies no longer need to install and maintain expensive and complex solutions (**challenge of high CAPEX**), a cloud option may seem very appealing from both an investment perspective, as well as in the long term. Not only do companies save directly by reducing setup and maintenance costs, the flexibility of a cloud solution often provides much greater indirect savings. For example, with a cloud solution, it is typically rather easy for companies to have employees working from home or remote locations—including in different cities or countries—where the cost of renting facilities may be lower.

There are two important aspects to consider when planning for a cloud-based contact center solution. The first is a fast and stable network connection. A cloud-based contact center is a system with components distributed across multiple locations, so connecting system components is an important prerequisite for overall success. The second aspect is planning where and how sensitive data will be stored and transferred. One option is to keep storing customer data on the premises while the contact center software and systems are in the cloud. This is often the preferred approach for banks and financial organizations. By keeping data on the premises, organizations avoid the challenge of using complex encryption mechanisms and addressing the complex regulatory compliance issues of customer data. Another solution may be keeping customer data in the cloud. This is often acceptable for small companies, especially when small companies don't have mature IT environments. For them, the cloud is more reliable and secure than any IT environment they could afford to establish on the premises. Companies may also choose a hybrid approach in which the data is partially stored on the cloud and partially on the premises.

Last, but not least, there are often options available for a mixed approach in which both systems and data are partially hosted in the cloud and partially on the premises. For example, a voice recording system is often hosted on the premises because it normally generates high volumes of data, and uploading that data to the cloud can be bandwidth-consuming. Another component often hosted on the premises is the voice gateway; because many companies already have phone lines installed at their offices and would like to keep it that way as well as keep the agents located at the same place. Sometimes, government regulations play a role in system design as some governments (and sometimes even internal company policies) dictate that customer data may not be stored outside of a company's physical premises. While components such as IVR and ACD do not normally contain customer data, components such as voice recordings or a historical reporting database can be very sensitive, dictating the choice to use a mixed design approach.

Knowledge Base

A knowledge base (KB) is, perhaps, one of the longest-standing components of contact centers, but it has been largely affected by the digital transformation and the shift from call center to contact center. As mentioned before, today's contact centers make thousands of contacts on a daily basis. You may observe that despite the growth in size, issues being served in contact centers are not unique. The Pareto principle—also known as the 80/20 rule—is applicable here. Though exact numbers may differ, the assumption that 80% of the contacts received is attributed to 20% of the potential issues is typically correct. This is why building a consistent source of information with documented solutions for frequently faced issues can make it easier for agents to quickly assist customers with typical problems, thus decreasing the AHT and increasing the FCR. Correct and appropriate business processing information needs to be at the agents' fingertips. A knowledge base is the universe of information required by agents to guide themselves and the customer.

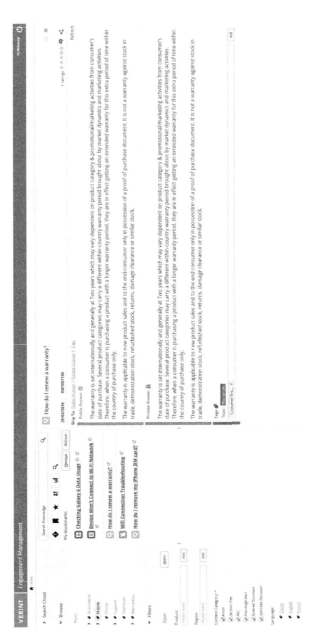

Verint Knowledge Management system suggests to the agent the resolution for the common customer inquiry about the phone warranty. © 2020 Verint Systems Inc. All Rights Reserved Worldwide. Used with permission.

A knowledge base is that source of information which, when built, configured, and properly administered, can make a huge difference in the way a contact center works. Agents can instantly search and find answers with the help of advanced search tools and AI. Even complex issues can be resolved easily and quickly. Let's look at the best practices applicable to the creation and deployment of knowledge base systems.

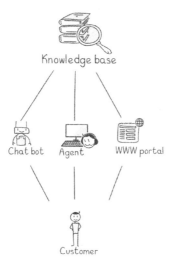

The first rule is that a knowledge base should be **centralized across all channels**. This makes the knowledge consistent and easier to administer (not to mention that one knowledge base product for all channels is technically easier to maintain than several KB systems). Customers often try to solve problems on their own before contacting contact centers, so it is better if the answers they get from different channels are consistent. Making a KB available via a company's self-help portal or integrating it with your chatbot solution will help customers to solve their problems without the need to contact the company seeking an agent's support.

The second rule, which should seem obvious by now, is that the agents should have **integrated access** to the KB as part of their desktop. A knowledge base will make problem resolution faster itself, but why should an agent spend extra seconds switching to the KB application?

The third rule is based on a fact already mentioned above: many customers will try to solve problems on their own before contacting an agent. If customers are using a portal to search for solutions and then click on the "chat with an agent" button or using the chatbot to get help before escalating it to a live agent or ask for a callback, then the agents should have a history of what customers have already seen and tried on their own.

Potentially, the most powerful approach is to use artificial intelligence to process information and suggest the best answers whenever possible. There is a great deal of information we might gather about customers from the CRM and previous interactions the customer has had with the contact center, ticketing system, and portal browsing history. We can also get real-time data about a customer's present interactions. For example, if a customer calls a power distribution company, we can identify the customer by his mobile number, get the customer's location from the CRM, and quickly retrieve the information that there is planned maintenance in the area and the power will be restored within 30 minutes. Then, we could pop-up a knowledge base article for the agent to tell the customer what to do for the next half-hour until the power has been restored (the answer will probably start with "keep calm"). Another example can be the contact center of the food and drug authority (FDA), which may be able to quickly suggest first aid procedures in case the wrong dose of medicine or the wrong medication has been taken by a mistake. With the current evolution of search and AI engines, the trend is toward self-service channels—such as chatbots and portals—replacing agents for the most common issues. This is only expected to continue in the future, completely changing the way customer service works for us.

Summary and Quiz

In this chapter, we discussed some of the latest trends and technologies which will drive the contact center and customer care industry in the near future. The digital world is changing faster than ever before; therefore, I suggest the reader complete more research on the topics discussed in this chapter as well as on general topics of digital and cloud transformation in the customer care industry, as the amount and speed of innovation out there is greater than I could realistically reflect in this short book.

Please answer the following self-check questions to test your knowledge of the topics covered in this chapter:

1. Which term describes the phenomenon in which people do the same things they used to do, but they do them differently due to digital technology:
   a. Digital acceleration
   b. Computer transformation
   c. Digital transformation
   d. Digital experience
   e. Technological transformation

2. Please choose two (2) important criteria supporting the use case of RPA for task automation (choose the two (2) best answers):
   a. Tasks must be repetitive.
   b. Tasks must only involve well-known applications (such as Windows, Office, etc.).
   c. Tasks must take a significant amount of time to complete, usually more than 30 minutes.
   d. Tasks must have a well-defined logic without complex decision making.
   e. Tasks must have associated service levels.
   f. Tasks must not include internal intranet portals as these are not accessible online.

3. An insurance company gets a few thousand motor insurance claims a day. A few forms must be filled out and submitted by employees for each claim, the accident report must be obtained online from the traffic police, and then the insurance expert reviews the claim and makes the decision. Afterward, the request must be sent to process the insurance payment for the customer or the work order for the car workshop to fix the car. Can an RPA help this insurance company to enhance the process (choose the best answer)?
   a. No, an RPA cannot be used because the process involves complex human decision-making.

b.   Yes, an RPA can be used to make the entire process faster as the process is repetitive and simple.
c.   Yes, an RPA can be used for all parts of the process except those involving external systems (getting traffic police reports online and sending work orders to fix the car). The work involving external systems must be done by humans.
d.   Yes, an RPA can be used for all parts of the above process except for the claim review.

4. Which of the following correctly describes the concept of a back-office workload distribution system?
a.   It is a component of a CRM system, applying a service level for each task, and controlling the execution step-by-step.
b.   It is a system distributing different tasks to employees based on their skills and load and ensuring the tasks are fulfilled within the SLA.
c.   It is an intelligent reporting system controlling the execution of tasks by back-office employees and providing valuable information about employee productivity to the management.
d.   It is a part of a contact center system, responsible for controlling tasks the agents perform during the after-call-work (ACW).

5. Which of the following are benefits of using a workload distribution system (choose all that apply)?
a.   It achieves the complete automation of tasks, thus reducing operating costs.
b.   It replaces traditional systems—such as CRM, helpdesk, etc.—with one unified system.
c.   It gives the management a clear view of the company's execution processes, helping to spot delays and take corrective action quickly.
d.   It ensures that tasks are picked and executed according to the company's business priorities rather than the employees' personal choices.
e.   It controls that tasks are executed on time and can be re-prioritized if they are about to break the service level.

f. It does not replace traditionally-used systems—such as CRM, helpdesk, etc.—but instead, it integrates with them.

6. Which of the following are two (2) important aspects to keep in mind when planning to switch to a cloud contact center (choose two (2))?

   a. It is typically impossible to integrate with on the premises telephony systems as the cloud uses its proprietary telephony protocols.
   b. A stable network connection is an important prerequisite.
   c. The CTI will be limited to what is provided out-of-the-box by the cloud provider and usually cannot be modified.
   d. Storage strategies for customers' sensitive data must be well-planned, and applicable government regulations should be considered.
   e. A high CAPEX of initial cloud activation must be considered.
   f. It is impossible to implement IVR services as the IVR system is located on the cloud, and backend systems are located on the premises.

7. Which three (3) challenges does cloud-based contact center approach solves:

   a. CAPEX is avoided or reduced.
   b. Integration with on the premises systems is easier and quicker.
   c. It is quicker and easier to scale the system up and down as the business requires.
   d. Cloud contact center systems are more flexible to configure per the company's business needs.
   e. Upgrades are faster and seamless as they are managed by the cloud contact center's solution provider.
   f. Licenses are no longer perpetual, but they can be subscription-based.

8. You are changing your company's large contact center from on the premises to a cloud solution. You are planning to keep the agents'

location in the same place. Please choose two (2) correct statements from the ones below:

    a. I will inform the management that the phone number needs to be changed to the one supplied by the new cloud contact center's solution provider.

    b. I will consider utilizing existing voice gateway and telco connections I already have at my location.

    c. I prefer to keep the voice recording system on the premises to store customers' sensitive information (calls) locally to comply with government regulations.

    d. I will ask the cloud contact center's solution provider to allocate additional space to store voice recording files and customer data on the cloud and ask my management to apply for a high-bandwidth connection.

9. What is the best practice for having the knowledgebase (choose one best answer)?

    a. Must be a part of a chatbot solution so that customers may use the chatbot to get help without live agent interaction.

    b. Must be a part of a contact center solution so that it can be utilized by all agents.

    c. Must be a part of an online web-portal so that customers may use it to get help by themselves.

    d. Must be a separate solution integrated and available on all channels (contact center, chatbot, and web portal).

    e. Must be on the cloud to be always available for customers, even if all channels are down due to maintenance or failure.

    f. Separate knowledgebases should be deployed for each channel (contact center, chatbot, and web-portal) because each channel deals with specific customer problems.

10. You are hired as a customer experience manager in a new, fast-growing digital bank, and the management asks you about the right approach for deploying customer care channels. Please choose the best answer:

a. Use on the premises deployment and subscription-based licensing to minimize CAPEX.
b. Use cloud-deployment to minimize CAPEX and ensure that quick business growth can be addressed.
c. Use hybrid deployment in a cloud-based contact center to minimize CAPEX and ensure that quick business growth can be addressed but store sensitive customer data on-site to meet government regulations.
d. Use on the premises deployment and perpetual licensing to minimize operating costs and have a solid system for years to come.

Bonus: All-in-one Diagram for Customer Care and Experience Technologies

I decided that representing the entire ecosystem of customer experience technologies in one neat summary slide would be beneficial for readers, to sum up everything they have learned by reading this book. Luckily, this complex task has been already accomplished by my manager, Mohamed Fahmy, whom I would like to thank for the permission to publish it here:

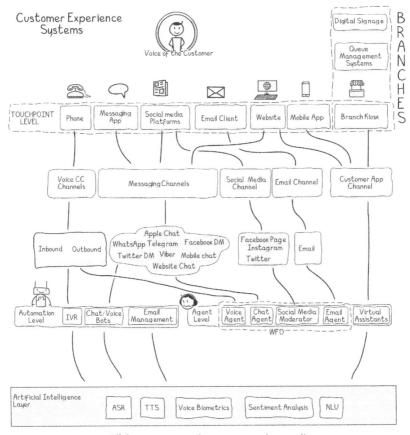

All Customer experience systems in one diagram

List of Acronyms

ACD—Automatic Call Distributor
ACW—After Call Work (also known as wrap up time)
AHT—Average Handle Time
AI—Artificial Intelligence
ASR—Automatic Speech Recognition
ATT—Average Talk Time
AQM—Automatic (automated) Quality Management
BHCA—Busy Hour Call Attempts
CAPEX—Capital Expenditure
CO—Central Office
CFM—Customer Feedback Management
CPE—Customer Premises Equipment
CR—Compliance Recording
CRM—Customer Relationship Management
CSAT—Customer Satisfaction
CTI—Computer Telephony Integration
CX—Customer Experience
DID—Direct Inward Dialing
DNC—Do Not Call List
DOD—Direct Outward Dialing (Direct Outbound Dialing)
DPA—Desktop Process Automation
DT (DX)—Digital Transformation
DTMF—Dual-tone Multifrequency
FCR—First Call Resolution
HR—Historical Reporting
IVR—Interactive Voice Response
KB—Knowledgebase
KPI—Key Performance Indicator
MOH—Music on Hold
NPS—Net Promoter Score
NLU—Natural Language Understanding
OPEX—Operational Expenditure
PABX—Private Automatic Branch Exchange
QM—Quality Management
QoS—Quality of Service

PBX—Private Branch Exchange
RPA—Robotic Process Automation
SA—Speech Analytics
SL (SLA)—Service Level (Service Level Agreement)
TTS—Text To Speech
VB—Voice Biometrics
VoC—Voice of Customer
WDS—Workload Distribution System
WFM—Workforce Management
WFO—Workforce Optimization

Answers to Quiz Questions

Chapter 1
1a. 2b. 3ade. 4c. 5d. 6b. 7b. 8b. 9c. 10e.

Chapter 2
1a. 2b. 3ade. 4e. 5b. 6c. 7a. 8d. 9b. 10abcd.

Chapter 3
1c. 2b. 3a. 4ac. 5ad. 6abc. 7acd. 8d. 9bc. 10c.

Chapter 4
1d. 2abde. 3c. 4b. 5ad. 6b. 7ade. 8abcde. 9abcd. 10bc.

Chapter 5
1c. 2d. 3d. 4e. 5abcd. 6c. 7abd. 8ac. 9b. 10bce.

Chapter 6
1c. 2b. 3e. 4a. 5ae. 6abcd. 7a. 8acf. 9c. 10afh.

Chapter 7
1c. 2ad. 3d. 4b. 5cdef. 6bd. 7ace. 8bc. 9d. 10c.

Printed in Great Britain
by Amazon

81542872R00092